T0324192

In Search of a Simple Introduction to Communication

Nimrod Bar-Am

In Search of a Simple Introduction to Communication

Nimrod Bar-Am
Department of Communication
Sapir College
M.P. Hof Ashkelon
Israel

ISBN 978-3-319-25623-8 ISBN 978-3-319-25625-2 (eBook)
DOI 10.1007/978-3-319-25625-2

Library of Congress Control Number: 2015960772

Printed on acid-free paper

This Springer imprint is published by SpringerNature
The registered company is Springer International Publishing AG Switzerland

...the best material model for a cat is another, or preferably the same cat.

Arturo Rosenblueth and Norbert Wiener
The Role of Models in Science (1945)
Philosophy of Science, 12(4), 320.

Acknowledgements

This book was written during the first two years of life of my son, Orian Shalom Bar-Am. From a researcher who devotes the bulk of his days and nights to the leisurely and carefree contemplation of more or less obscure theoretical problems, I became an elated father, proud and bleary-eyed, who barely manages to squeeze a single hour of work into a day filled with prosaic child-rearing tasks. I don't know what and how much of this state of sensory intoxication found its way into the present book. Maybe only very little. It would be easy to blame this on my pedantry as a writer and rigid training as a researcher. But I think it is to the presence of my son that I owe the force of my desire to lay down thoughts that will be clear to all, and with the least possible effort on the part of lay readers, whatever the cost in slowing down my writing. I am grateful to you, Orian, for your bedazzling presence. That I was nonetheless able to find the necessary time to put my thoughts to paper is a privilege I owe to the generosity and love of my wife, who is so dear to me, Gali Drucker Bar-Am. Without you, this book would not have come into this world, to say nothing of our treasure.

Like much of my intellectual work, this book is a summary of conversations with my teacher, Joseph Agassi. Our conversations began more than two decades ago when I was his student and research assistant, and they continued and evolved very organically when I became an independent researcher and colleague. Throughout these years, Agassi has given me hours upon hours of his valuable time and abundant wisdom with a generosity whose limits I have yet to find, even though I have often tried. I am deeply grateful to him for providing me, almost offhandedly, with such a lofty and inspiring model of the love of the labor.

Special thanks to Robert S. Cohen for his support and invaluable patronage. Thanks also to my colleagues Ian Jarvie and Raphael Sassower for essential criticism and encouragement.

I am grateful to my students past and present, and especially to those of them who had the guts to admit that they don't understand, and to ask for explanations.

A special note of thanks to Natalie Melzer, a craftsman of the written word, who is highly adept at the task of understanding the other. Without her, the English edition of this book would have gone to print with what to me would have been an intolerable delay, and it would probably have been much paler.

Finally, I wish to thank the wonderful employees of Betta Café in Ramat Aviv, and especially the always-gracious shift manager Ofer Eisbruch, for providing me with a warm and pleasant second home and with unbeatable coffee once my work nook at home had been transformed into a joyous playground where serious-mindedness no longer belongs.

Contents

Part I
Getting Acquainted

Chapter 1
Who Am I? Where Do I Come From? Where Am I Headed?

Allow me to introduce myself. This is not just a matter of courtesy. It will help me write this book, and it will help you read it (or else perhaps put it down in favor of a different book). My name is Nimrod. I am a philosopher of science by training and a lecturer in a communication studies program by profession. In my classes, I teach the foundations of the theory of logic and of critical thinking. In addition, I teach and develop the field of study with which this book is concerned: the philosophy of communication.

Philosophers, so I have been told since my earliest days as a student, seek to know a little bit about everything: they hope to introduce a measure of order into the chaos of the world, that is, to orient themselves in our miraculous environment by examining general explanations for the occurrences they find within it. Were it possible, they would surely have wanted to read the book that is an introduction to all things. This kind of wish strikes many people as rather odd or naïve, especially today, in the age of information and of ever branching-out specialties. Not a few of the people I encounter respond to it with a chuckle of the kind reserved for harmless eccentrics. Without pausing here to challenge this dismissive attitude in any depth (It seems to me that some of history's most useful and most harmful, and in any case most influential ideas resulted from just this kind of "eccentric" thought), I want to emphasize that the debate about the conditions for successful orientation within our world is a juncture that we will come upon at many turns throughout this book, sometimes precisely when we seek to distance ourselves from it as much as possible, in an effort to be practical. This is because successfully orienting ourselves in the world is not just the eccentric ambition of the philosopher qua philosopher, but is of course the hope of every human being as such and indeed the aim of every organism insofar as it is alive—for life itself is enabled by, and depends upon, a proper response to the environment. Even the air conditioning system cooling my room as I write these words self-adjusts its level to the temperature in the room and thus performs a rather impressive task of orientation in its environment. As we will see later on (perhaps to the surprise of certain experts), this wondrous ability—*to orient ourselves in the environment*—is the most basic phenomenon that communication researchers seek to understand. The attempt to comprehend it yields fascinating and surprising puzzles. I suggest, and this is the main thesis I advance in this book, that the search for a general answer to the question of how this

N. Bar-Am, *In Search of a Simple Introduction to Communication*,
DOI 10.1007/978-3-319-25625-2_1

orientation successfully occurs is the vital element of the many varied attempts to understand communication in the conventional sense of this term—that is, our ability to share narratives and challenges with individuals who are similar to us biologically, culturally, biographically and so on. To use terms that we will gradually come to apply here in a highly specific sense, my claim is that *communication scholars study the ability of various systems to share environments*. Through this ability, such systems typically become parts of greater systems, with new and sometimes surprising environments. And the study of this process, the study of the creation, calibration and coordination of environments, is the study of communication.

There is no doubt something ridiculously pretentious, but also quotidian and obvious, in our desire to get to know our environment, to orient ourselves within it (and even in our partial success in doing so, at least relative to certain objectives). Philosophers differ from specialists and from laypeople (and of course also from simple organisms and air conditioners) only in the sense that their search for orientation in our world is grounded in a constant, conscious attempt to enhance awareness to the overwhelming magnitude of this simple-looking task, to caution us about its various pitfalls: they help articulate and internalize a set of precautions against such pitfalls. It is naturally much easier to be led astray by faulty principles if you have never asked yourself what these principles are, how they are apt to mislead you, and how, if at all, their harmful effect can be avoided or reduced. I offer this not as a dry academic point. Here you are, consulting a book that presumes to be an introduction of sorts to communication studies; your goal is probably to get an overview, a birds-eye perspective, of a vast and confusingly rich domain; you want to get to know it without becoming bogged down by too many details; is success in such a mission even possible? If it is not, how can your objective be fulfilled? Maybe your wish is to roam the vast expanses of this domain more or less causally without necessarily learning its general map. Many readers indeed prefer this way of acquainting themselves with new content like a traveler who has wandered into an immense deserted orchard, these readers stroll about idly, tasting whatever fruit strikes them as appetizing, perhaps in order later to return and eat at length and in a leisurely fashion of the fruit of those trees that were most pleasant to their palate. Even those who proceed in this way still apply, consciously or unconsciously, certain general habits of roaming, navigation, and thus of orientation that are at the same time also habits of learning. Whether these habits culminate in the drawing of a detailed map or, rather, in the semi-random marking of places worth revisiting in the future, they express a desire to organize and regularize future expectations in the face of what appears from a distance to be a complex, highly detailed space, or at least a vague and blurry mass. How can such tasks be carried out successfully? How do we all do this, every single day, every hour, and not just with our minds but with every part of our body, with every one of the cells that make up our organs and limbs, and even through the delicate coordination between their smallest components? Why do they do this?

My own professional life has been devoted to studying the wonder known as communication. True to my philosophical training, I have pursued this task from as general a perspective as my education and knowledge allowed me. In my Master's thesis, for example, I surveyed the debate about the question of how, if at all, do languages limit what can be said (and thought) within them. My doctoral thesis was on the question of what we learn from logic that collection of abstracted inferences perceived, from Aristotle through Russell, as "the mother of all languages". All this happened quite a while ago. I mention it here as an example of the level of abstraction, or of generality, that a philosopher seeks as an inseparable part of his education, even if he is a young and not exactly highly experienced researcher (is it possible to make real progress toward valuable goals in this way? How so?).

Roughly a decade ago I found myself teaching in a communication studies department. It was my first close exposure to this academic field. I met new fellow researchers: experts who had trained from the start as academic communication researchers, and others who, like me, were trained in other disciplines and ended up, to their surprise, working at the heart of this fascinating field. This new encounter provoked in me enthusiasm, but also a certain uneasy confusion. As I said earlier, I had arrived in the department as someone who already considered himself a researcher of communication from a rather general and abstract perspective. To my surprise, there was very little in common between the problems that I was trained to study and those that preoccupied my new colleagues. These colleagues had actually gathered in communication studies from all corners of the academic world, and as far as I could judge at first glance, they investigated communication from perspectives that were diverse to the point of chaos: there were among them sociologists and anthropologists who studied the communication between and within communities, historians who studied the development of various technologies and their contribution to the evolution of humankind, behavioral economists who studied our potential to understand and influence consumer decision making, psychologists and linguists who studied interpersonal communication and group dynamics alongside psychologists who studied human perception and the neurobiological mechanisms that enable it; their work relied on the research of physiologists, ethologists and sociobiologists who study animal communication and the evolutionary origins of orientation patterns, and to various degrees, they all kept abreast of the work of software engineers and computer researchers on machine communication and the communication among computers in particular. This varied lineup of researchers attempted to provide theoretical support to another group, no less confusingly de-centralized. This was a team of more or less renowned figures who were undoubtedly the main source of the department's attractiveness to students—current or former media professionals who led alluring workshops in various areas in which they had real-life, hands-on experience, primarily (but not exclusively) in the areas of practice grouped under the misleadingly definitive title of "mass media": writers and reporters, editors, photographers, advertisers, radio and television personalities, public relations and sales professionals, entrepreneurship advisers, internet experts, etc.

What struck me above all else was this: how little communication actually took place among the members of this heterogeneous group. My new colleagues did not appear to share anything resembling a distinct discipline. More importantly (since

I am no great believer in disciplines and do not consider their absence to be a particularly serious deficiency), they did not appear to share even a tentative list of basic (or even just traditionally accepted) problems. Now this deficiency is indeed significant because, as I will later argue, the ability to share problems is the single most important pre-condition of communication. It is not surprising, therefore, that my colleagues also had no common toolkit, or anything approximating one, with which to approach the task of solving the wide-ranging problems they tackled. Indeed, as a result of these deficiencies, they had no common professional language—with the exception, as in every specialized activity, of several jargon terms that served mostly to demonstrate the writer's awareness of and/or loyalty to one or another academic faction. No wonder, then, that the possibility of communication among these researchers was severely limited. Only one observation, so it seemed to me, emerged as undeniable and widely accepted among members of the field: it seemed to all of us that, for reasons that were hard to identify, the communication between scientists and the communication between computers (which in many ways represent the two most remote poles of our field of study) have more in common than the sociology of science and computer science (which are, more or less, the disciplines that study these phenomena). In other words, we all agreed that there is room—indeed, a need—for a discipline that gathers phenomena from various disparate disciplines and studies them as instances of communication.

Because I was still new to this academic field, I tried to get my colleagues to explain to me this chaotic state of affairs. Where was the general introduction to the theory of communication, I asked. Where could I learn something about the field's general problems, i.e., those problems common to the research of communication in all of its forms? (Later on, I will address the question of the distinction between the research of communication-in-all-of-its-forms and standard communication research as it has come to be defined by the schools of communication.) Where would I find the central controversies that define our field? Did there exist anything like a set of guiding principles (however rough or tentative) that might help me put some order in the apparent bedlam of research into which I had stumbled? For the most part, my questions were met with a shrug accompanied by a statement that ran contrary to everything I had known up to that moment: I was told that since communication (in the most general sense of the word) concerns phenomena that range from the domains of physics and computer science to sociology, history, and economics, the field was in any case too broad and decentralized for the researchers investigating its far-flung corners to share any common problems. This broadness and decentralization, I was told, are the reason for the ultimate absence of any real introduction to the field… the simple introduction to communication-in-all-of-its-forms.

This statement struck me at first as a curious absurdity. It went against everything I had been taught as a researcher: after all, the fragmentary or decentralized nature of a field of study is just further evidence that we still lack the general principles that will allow us to orient ourselves within it. It has nothing to do with the field's broadness or scope. Consider the size of a continent, for example; in itself, size poses no barrier to our ability to map the land. We need only to find the tools that will carry us high enough above ground so as to observe it properly from what we call the "bird's eye view". Then it will no longer be a seemingly random collection of fragmentary data about a lake, field, or mountain that we just happened to come upon, but rather a full-blown map. Is it possible that a theory that would encompass various different systems of communication poses a challenge of a different order? Could it be that communication is actually unlike continents insofar as we cannot, as a matter of principle, fly high enough above it? And if so, why? (We map even spaces above which we cannot fly in actual practice, such as the universe; is the analogy between knowledge and cartography stretched here beyond its legitimate limits and therefore unintentionally leading us astray?) If this is the case, I thought to myself, then we should at the very least articulate to the best of our ability *why* this is so, and what can be done to attain even a partial understanding of this strange and decentralized space, which is so fascinating. My colleagues and I, as I said, identified various kinds of phenomena as clear-cut cases of communication. Consider, for example, the remarkable fact that readers of English now reading this page are in an environment radically different from that of non-readers of English who face the same physical presence, namely the printed page: Is this not a truly extraordinary fact? How can it be explained in physical terms? My colleagues and I distinguished between phenomena of this sort and other phenomena that are not, in our view, distinctly part of our field (such as the study of the chemical composition of the page). Is there a theory that comes close to uniting and resolving, even roughly, the puzzles described above as distinctly related to communication? "If you are right", I replied to my colleagues, "if communication is indeed too broad and decentralized a field to be abstracted and brought under a single, systematic theory, then computer communication and the communication between scientists have no more in common than physics and sociology… what would be the fate of our aspiration to understand communication-in-all-of-its-forms, then? Moreover, what, if anything, can we offer the student who seeks such an understanding? For we have just called into question the very right of communication studies to exist as a distinct field of research".

As soon as we conclude the introductory chapters, we will return to this problem in as much detail as the scope of the debate in this brief book allows. It forms an inevitable starting point for every study in communication that presumes to be comprehensive. The issue is all too often sidestepped, hastily and quietly, in standard introductions to the field. Indeed, the purpose of this book is to re-kindle open discussion of the desirability of such hasty and silent sidesteps. The challenge we face here is immense, and also, whether we like it or not, at once highly abstract and practical: the term "communication" normally denotes interesting aspects of almost every familiar fact, and encompasses fascinating problems in everything from physics and engineering, biology and methodology, to sociology, economics, and history. The theory of communication, were we to find such a thing, seems at first glance like a general theory of almost everything there is, or more precisely, a theory that tackles some of the prominent difficulties that every theory expressing such a terrifying (or ridiculous) and brave (or contemptible) pretension must tackle.

It is therefore imperative that we try to separate the ridiculous from the valuable. This sort of debate will lead us inevitably into investigations of a general and sometimes metaphysical nature. We will need to consider at least the merit of developing such a general theory, and its potential value to the researcher, the student, and the media professional.

As part of our initial acquaintance I should also add that in my search for the fundamental problems of communication research I surveyed every introduction to communication studies I could lay my hands on. And almost every time, I found myself disappointed. This fact is noteworthy, and not for what it may say about me: the vast majority of books that presume to offer general introductions to communication studies seem to mislead their readers. In the best (and rare) cases, the authors fail to align their declared goal (to provide a theoretical introduction to the field of communication studies) with their actual practice, which is a description of interesting chapters in the history of various technologies accompanied by some form of survey of theories from the social sciences. In the less successful (and more common) cases, the authors actually do their readers harm, in my view, by offering them, in the guise of an impartial introduction to the field, an initiation into one or another academic faction to which the author belongs and which amounts to little more than a united group of academics whose studies are an expression of their tribal loyalty to a theoretical corpus of common forefathers. In my experience, it is only rarely that students who have been introduced or inducted into a field in this manner succeed later on in finding their way back to a critical, independent, sober perspective.

Contrary to all this, shining brightly in their solitude and sometimes also in their marginality, stand the pioneering studies of several of the field's founding figures. It is telling of the state of the field's research, I believe, that even today readers will find more value in the works of such scholars as Norbert Wiener or Marshal McLuhan than in many of the writings of the most popular current scholars who regard themselves as the contemporary successors or critics of these seminal figures. Other figures who have greatly influenced the thoughts expressed in this book include the chemist and sociologist Michael Polanyi and the perception psychologist J.J. Gibson. Their work, however, is little known to communication scholars as such. I will make use of their influential ideas later on in the discussion, as this book is largely an attempt to re-examine their proposals, identify their limitations, moderate them where necessary, and consider the possibility of expanding them in light of the technological and intellectual revolutions of the past several decades. It is rather disheartening to discover that their writings, sometimes more than fifty years old, are still among the clearest attempts to articulate the basic challenges facing those who seek a general introduction to the field. This book, then, represents my own attempt to explore the possibility of updating their ideas wherever necessary and to document the difficulties I encounter in those cases where so doing was found to be beyond my ability.

One more interesting fact is worth mentioning as part of our acquaintance. Since I did not find what I wanted in the standard introductions to communication studies, I also read numerous less standard introductions. This exploration was useful to me in two senses. First, this expanded search drew my attention to the impressive circulation (and sometimes also considerable commercial success) of books that proclaim themselves to be introductions to communication studies but are nothing of the sort. I am referring to books boasting titles such as "Why Study Media?" or "Everything about Communication", whose contents fulfill not even the smallest fragment of these promises. From them I learned an important indirect characteristic of the field that I was struggling to understand: the field is a fertile ground for the hallucinations and visions of dubious scholars, decadent or vociferously romantic sociologists, and paranoid futurologists (or else ones who cynically exploit the widespread fear of modernity). The market, it appears, cannot get enough of such writers, just as it always craves more empty self-help guides. As a marketing expert once told me, soap will always cost much less than anti-aging cream, and for a simple reason: whereas it is easy to fulfill the soap manufacturer's promise to deliver a product that will cleanse your skin, and even to put this promise to an empirical test, the manufacturer of the anti-aging cream promises the consumer something that both seller and buyer, in their heart of hearts, know to be unattainable: to eliminate the signs of aging, and give us a more beautiful face. The inflated price of anti-aging products, then, is essentially a tax imposed by the manufacturer because consumers lack the courage to admit their own collusion in an act of willful self-deception. Thus, even from these less helpful introductions to communication, I gathered that I would need at least to address the questions of why it is that the field attracts so many curious minds; what claims (both empty and real) it promises to fulfill; and why it is so easy in this field to present oneself as the expert-of-the-moment while offering no more than a smattering of new-age techno-gothic ruminations atop a patchwork of the standard introduction to sociology or a medley of memoirs from a long career as the editor of a well-known long-gone publication.

The second and more significant sense in which the search for non-standard introductions helped me is that it introduced me to a rich world of fascinating attempts to grapple with the central difficulties of communication studies that just happened to lie outside the boundaries of the standard academic discipline. For this reason, many of the excellent traditional researchers of communication I have met may be insufficiently aware of these works. For example, the field known as "systems theory" (misleadingly, of course, as it designates no theory) encompasses a collection of critical attempts to tackle some of the central challenges involved in the articulation of a general theory of communication. But it has grown out of the work of researchers in such fields as engineering, physics, biology, and even psychiatry and meteorology. The formal study of communication is often far removed from these thinkers, and their conclusions need to be translated into the language and terminology familiar to students and teachers of communication. Such are also the many and radically different studies in ecology, evolutionary biology, and even the mathematical study of networks, and, in particular, studies by those researchers who sought to apply basic insights from these fields to the social sciences and the humanities. For instance, I am happy to acknowledge my own debt to the writings of Edward Reed, a perception psychologist (a devout disciple of the abovementioned J.J. Gibson) and an original philosopher of mind, for they have drawn my attention to some of the major problems I discuss in the following pages. I am also deeply indebted to the work of Reed's philosophy teacher, Joseph Agassi, who was also my own philosophy professor and mentor. I mention my sincere gratitude to these outstanding scholars now both because it is my pleasure to acknowledge their influence publicly and because in my attempts to make this book as readable and palatable as possible I have often avoided discussing in detail the sometimes highly intricate scholarly genealogy of my ideas.

The reason that the search for a theory of communication appears in so many versions and has grown out of such varied scientific pursuits will be explored in detail in the coming chapters. For now, let me just note that it is clear that these researchers, who hail from a wide range of disciplines and share nearly no terminological background, came upon similar difficulties in our modern scientific picture of the world, the picture on which they were all trained as students and later worked as academic researchers or professionals. The most fundamental of these difficulties concerns the fact that our comprehensive scientific picture of the world presents our world as uniting physical, psychological and sociological elements with no obvious bridge that connects them and only the persistent faith that one day such a bridge will be discovered in the form of a reduction of all the sciences to the language of physics. The wonderful dream of such a bridging (or reductive) translation, I will argue in the following pages, is moving farther away from us at the same frightening pace at which our understanding of the conditions for its success is progressing. The attempts of these thinkers to grapple with this difficulty were vitally important however, as we will see here, for they have advanced our understanding of the possibility and limitations of a general theory of communication. They pursued this task, typically, using the unique vocabulary and terminology with which each one of them was familiar and with special attention to those problems that are particular to their respective specialized fields. This sometimes caused an unfortunate divide between their works, one that truly calls out for some sort of agent or medium of mediation. It will be our task at least to get to know the basic problems tackled by these fascinating researchers and to comprehend why the attempt to generalize their solutions as far as possible in a common language—the language of those who seek to study communication—is a worthy effort.

In sum, what I aim to do in this book is to explore the best available attempts to sketch the outline of this vast and mysterious continent—the study of communication and the study of media—with all of its lands, forests, orchards and deserts, in the hope of succeeding at least to assess the general limitations of this sort of project. As you will see, I think there is a sense in which my shoulder-shrugging colleagues had it right: there are indeed valid and interesting reasons to qualify any theory that presumes to offer a comprehensive and strict explanation of the world of communication. This is, as we will see, perhaps the most important feature of the field: communication is not at all like a painter's nude model who stands still and waits patiently and passively as its noble geometrical contours are slowly traced by a calculated, expert draftsman. On the contrary, from the very first encounter with communication it is obvious that we must recognize the limitations of the worldview that treats scientific knowledge as a static drawing or map of this sort, and communication as the context-independent transmission of this drawing or map to others. Communication is a dynamic whose rules undergo change even while its members are in the process of studying them; it alters its members while they are sharing it, alters the communicators as they are communicating and the objectives of those who study it while they are still articulating them. This feature will emerge as a central challenge facing anyone who seeks to study communication. Students of communication, therefore, must first learn general instructions on how to pursue an outline that is in constant flux, its features shifting and changing during the course (and sometimes as a result) of their attempts and occasional partial successes to capture it. Happily for us, this kind of complicated task is not as unfeasible as it may first appear. Or in any case, it is not always so.

One of my teachers once told me that the shortest way to find a book that does not exist is to write it. *The present book, for better or worse, is a preliminary report on my search for the introductory book that I was never able to find: the simple introduction to communication studies. Its goal is to document for my students, in brief form, the difficulties facing anyone who wishes to study communication-in-all-of-its-forms.*

Chapter 2
Who Are You? Why Are You Here? Can I Help You Achieve Your Goals?

In the previous paragraph, I mentioned my students. It is thanks to them and for them that this book was written. I must say straight away that I owe them a debt of gratitude: the confusion I experienced as an old-new researcher in a field that lacks a systematic introduction was dwarfed by my bewilderment as a teacher in that field. For, in the classroom, the practical aspects of this problem were thrown into sharp relief and the challenge that they pose took on an added urgency: how could I be of value to such a varied group of students?

Seated before me in my classroom were: a future journalist and a future sociologist of the web, a future publicist, radio broadcaster, photographer, historian of technology, television anchor, publisher, political spokesperson, conflict resolution expert, advertiser, software engineer, web designer, marketer, management consultant, and even a future poet (even if still unbeknownst to her). The list of potential professions is long and rather open-ended, since we do not know what form media professions of the distant and not-so-distant future will assume. From the academic perspective, communication studies have quickly and surely replaced what was known until recent years as a "general BA", at least insofar as the same large section of the population consisting of those who seek an academic education but do not yet know what field they wish to study and who twenty years ago would have ended up in the default program of "general humanities" today enroll in communication studies, which are considered the practical general studies of the early twenty-first century (a century that no longer lets the inquisitive student roam freely and marvel at the treasures of human culture without the burden of defined financial objectives and pre-articulated career goals).

And yet, in the past decade there are signs again of changing winds in the behavior of this segment of the population: at least part of the large group of students that until recently had sought out its education in departments of communication studies has begun to move, somewhat uneasily, toward business and management schools, of all things. It is important to recognize that both the massive shift that began some sixty years ago toward communication studies and the shift of more recent years, which is accompanied by a certain sense of a crisis of faith, away from communication studies toward business and management schools, are a distinct symptom of the same sense of lack of direction that concerns us here: the lack of general and clear guidelines for understanding the sum of all phenomena

N. Bar-Am, *In Search of a Simple Introduction to Communication*,
DOI 10.1007/978-3-319-25625-2_2

described as "communication" is the source both of the heightened interest in the discipline and of the intense criticism to which it is subjected. (Whether or not those who throng to business programs are aware of this, the difficulties that brought them to abandon communication studies in favor of business departments rear their heads in those programs just as stubbornly and for the same reasons—because business administration is rather obviously just a minute sub-field of communication studies, at least insofar as it shares the same methodological difficulties and limitations that we will consider here.)

I wish to help the diverse group of students seated before me in class. I note to myself that the cluster of professions in which they hope to train is not as rich and complex as the selection of professions and training backgrounds of the teachers of communication, not to mention the selection that includes everyone worthy in principle, from an abstract and theoretical perspective, of being deemed a communication researcher—that is, anyone studying central aspects of communication-in-all-of-its-forms. What are the professional problems common to all of these areas of expertise? What are the qualifications and professional skills that characterize experts in these professions? Why have they been lumped together? The answers to these questions are not as clear-cut as some teachers in communication studies have become accustomed to telling themselves and their students. For example, it would be hard to find in our classroom a future professional linguist, even though the linguist is undoubtedly a researcher of what is our primary medium of communication—human language. When I suggest this in class, a student raises her hand and asks why the future educator and the future musicologist are also absent from our list of students. She is right, of course. These are indisputably important and central domains of communication. And we could easily go on to expand the list: it also lacks, among many others, the various future artists, who are of course all in the business of communication, as well as translators, philologists... and this, without even venturing into the so-called "harder" sciences. For neither do we normally find in courses that offer an introduction to communication studies the future computer scientist nor the future evolutionary developmental biologist, despite the fact that were there a course that offered a genuinely general introduction to communication, an introduction to communication-in-all-of-its-forms, I have no doubt that the students of these disciplines would take their places at the front of the classroom with an eager and urgent curiosity.

The confusion only increases when we notice the ambiguousness of the name by which the study of communication commonly goes: "media studies". The literal meaning of this deliberately vague term is "the study of means" ("media" is the Latin plural form of "medium" which means "middle" or "in-between"; a medium, then, is anything that stands in-between things). The literal meaning of the title "media studies", therefore, is the study of the mediators. Perceptive students will rightfully ask: *which* mediators are referred to by this title? For if everything that stands between two (or more) things is included in our discipline, we will find ourselves seeking a theory of everything that exists... Is it possible to limit our field so as to enable a responsible investigation of certain distinct parts of such a vast world? What are the forms of media we wish to study, and why? Is their list finite? And if not, is there some sort of criterion we can apply to these forms of media to distinguish those that interest us form those that do not? We seek to study only those media that function as means of communication, of course. So why drop the second half of the noun? Could the reason be that we simply do not know to distinguish a medium of communication from a medium that is not one of communication? Is there a medium that is not simultaneously a means of communication? Here colloquial English serves to cover up a confusion: those who named the field "media studies" appear to have preferred vagueness with respect to the term's designation, perhaps hoping that its users would somehow fill in the missing criterion. But the confusion does not go away by itself. This is attested by the countless unsuccessful elaborations of the term that can be found in media studies departments across the world and in discussions of media in the literature. One widely used term, for example, is "the technical media", and there is also "the new media". These elaborations are useless from a theoretical perspective because there is no medium at all that does not rely on techniques and no medium whose operation does not presuppose a skill (i.e., that is not "*techne*" in the original Greek sense of that word), and also because we would need to know, at least vaguely, what does and does not counts as media before we could determine if they are indeed "new" media, as opposed to variations on old media (or maybe, if this is possible, a new phenomenon that is not at all a medium). The confusion, then, remains in tact despite the attempt to conceal it behind pseudo elaborations and groundless distinctions, and the terms that describe the field slip under the radar of those who employ them as if their designation were perfectly clear, as if every intelligent person knows what the media are. Other well-known tags for sub-categories within our field express similar difficulties: for example, "Organizational Communication", "Social Communication", "Interpersonal Communication", "Cultural Communication", and even the recent "Media Ecology", among many others. Whether or not we are sympathetic to this illusory solution of brushing the confusion under the dictionary, we should not underestimate the difficulty it attempts to solve: the search for the unique element that unites the many fields grouped under the academic heading of media studies is nearly as tangled and complex as the search for the unique element that unites all the phenomena normally included in the category of communication-in-all-of-its-forms.

Some of the communication studies experts I met over the years seemed to have internalized, perhaps as early as their student years, a set of institutional habits and patterns of thought that allow them to brush aside the doubts I raise here with a confident wave of the hand, or at most a slight rise in intonation designed to drown out undesired questions. The field taught in communication departments, their dismissive explanation goes, is of course not the product of the study of communication-in-all-of-its-forms, if such a study even exists, but merely the product of the study of *mass* media. It is a historical fact that certain technologies, like speech and writing, have shaped the development of humankind, and that some of these technologies, like print and the telegraph, have had a decisive effect upon the evolution of modern society and therefore also on our becoming a multitude, a mass (which is, of course, one of the prominent marks of modernity). Given that a complicated reciprocal relation holds between the cultural structure of a society (including the various institutions that cradle and establish it) and the psychic structure of its members, a relation that we will later dwell on at considerable length, the goal of mass media (they contend) is simply to understand the psyche of modern individuals by understanding the process of their formation by these communication technologies. The study of mass media, then, according to this view, is the interdisciplinary core of the social sciences: its students seek to familiarize themselves with a range of research areas beginning with the psychology of modern individuals and culminating with the study of the sociology, history, and economics that have shaped them.

This answer is correct as far as it goes. Or else, at least it faithfully describes the theoretical outlook of many departments of media and communication studies. In some respects, it is also historically partially accurate, that is to say, it faithfully captures some of the intentions of the field's founders and in particular of those who came to it from the humanities and the social sciences, like the brilliant economist Harold Innis and the unconventional professor of literature Marshall McLuhan. Yet I think it is important to pause over this answer and note how little it actually does to advance a theoretical introductory discussion of the sort that we are here pursuing: it replaces genuine confusion with an illusory security. After all, in the previous passage we found ourselves talking about the "mass media" and about "communication", while in fact we possess not even the beginning of a distinction that would allow us to identify and understand the meaning of "communication" and of "media". What sets mass media technologies apart from other technologies, including those that can be called mass technologies? Indeed, is there any technology that is not simultaneously a medium? And if not, where does mass communication end and another, more intimate form of communication begin?

In order to clarify this point it is helpful to pause briefly over the working definition provided by McLuhan (who was, incidentally, among the first and most influential thinkers to wield the term "media" in a deliberately loose manner as a characteristic label of the field). *McLuhan famously argued that a medium (a means of communication) is any object, material or theoretical, that we use to enhance a certain aspect of our being.* Take a stone, for example. The stone, argued McLuhan, if properly held, functions as a tougher, stronger fist, or a sharper claw, insofar as the man who picks it up while hunting, warring or working, effectively uses it as an enhancement of his hand. As such, the stone is a means of communication according to McLuhan. For using the stone *reshapes the environment* of the individual who picks it up (it makes it softer, for example) and will consequently eventually *reshape its user*. The users' hands, for example, no longer needing to be as hard as the rock, will soften too. The users' gripping skills will improve, and their brains will have evolved, too, so as to allow this. Moreover: when properly cast at its target, the stone is not just a harder fist but also a longer arm, and in this sense, too, argues McLuhan, it is a medium. For it shrinks the world of reachable objects. Finally, since stones have a much longer lifespan than people, they can be used to express various intentions (for instance, through a stone monument) that will echo for thousands of years after the passing of their creators. Thus, a stone monument represents a significant enhancement of our ability to deliver thoughts through time and even to enrich future generations with our experiences. Similarly, the automobile is an enhancement of our legs and movement capabilities, as well as of our ability to manifest social success; the library—an enhancement of our memory; the phone—of our voice; the binoculars and microscope—of our eyes; and even the economic system is a more efficient (and under very particular conditions, also more moral) arrangement of our cooperation. These, according to McLuhan, are all (communication) media.

McLuhan further argued that the change effected upon us by every such enhancement is total: it instigates a chain reaction that reshapes our identity by rearranging the relations between the communication means that we comprise. This observation is as trivial as it is profound: by enhancing any one aspect of our being, we simultaneously make certain other aspects redundant and rearrange the interrelations between all aspects. Clearly, the Giraffe's neck could not have grown to its present size without many corresponding changes in its ancestor's structure, brain and behavior patterns. Think, for example, how an extra pair of hands would influence our existence, or what would happen to us if we suddenly acquired eyes in the back of our heads. Such drastic biological changes could of course never occur in us suddenly, out of the blue, because they cannot occur without comprehensive corresponding changes in our brain, our spine, our sense organs, and all the rest. But our artificial enhancements, our discoveries and technologies, do occur quite suddenly. And their effect on us is often nearly as total. Like an extra-pair of eyes, artificial enhancements reshape everything about us, including our physique, our habitats, our psyche and the very structure of our society. Changes almost as radical as this did indeed occur in our ancestors, for example when they acquired the skill to kindle and control fire or when they began communicating by means of a rudimentary language. And recently, such changes are occurring at an ever-accelerating, often terrifying pace. Observing all of this led McLuhan to coin his best-known slogan: "The medium is the message". The changes effected upon us by new media, he argued, far exceed the changes caused directly by their capacity as vehicles of messages. This is true not only with respect to the discovery of literacy: it is true of print, of the telegraph, of the modern car, of the electric bulb, of the internet and the mobile phone, and indeed of every possible technology that we can imagine. Media are created by us, and yet they reshape us, often quite radically and unexpectedly.

McLuhan's insights are fascinating and illuminating. But for the purpose of the present discussion they serve primarily to underscore the confusion of those who, like ourselves, are wandering around a new domain in search of its boundaries or at least of some of its prominent features: not only do these insights fail to help us sketch the outline of what ought and ought not be included in the study of the means of communication, they also clarify how hopeless such an attempt would be in the first place. Ask yourselves whether there is any human technology—from a cotton ear swab or a napkin to the atomic bomb or particle accelerator—that is not a means of communication according to McLuhan's definition. A definition of this sort, therefore, is virtually useless as a tool for circumscribing our field: it effectively pronounces all actual and theoretical objects as means of communication. Could it be that those who wish to study communication are seeking a general theory of everything?

McLuhan, as it happens, was among the first to recognize the bulimic nature of his own definition, even acknowledging this fact with the kind of palpable amusement that came to be a trademark of his playful writing: he embraced the totality of his new theory with a wink, like a mischievous boy who stumbles upon an endless sandbox covered with toys as far as the eye can see. This is also the reason why he encouraged the use of the vague term "media": he wanted no boundaries on the study that he had helped to establish. In doing so, he distanced himself from those who regard the search for a general theory of everything as a case of shameless charlatanism (however charming, witty and original its presentation). Suppose, for example, that someone wants to study the history of media. According to McLuhan, what such a person seeks is the story of every technology (actual or theoretical) ever created, and of every kind of impact such technology has had on its users. Such a total history (the term "total history" was coined disparagingly by professional historians to denounce such pretension as ridiculous) is fantastic in the manner of Borges' endless, absurd library: we cannot even come close to producing its index, even if the whole of humankind applied itself to this task until the end of time. True, certain responsible and cautious scholars, like the abovementioned Harold Innis conducted meticulous studies of the history of certain basic goods (such as fur) within circumscribed geographical areas, as well as of particular technologies (the railroad track, for example) and of some of the ways in which they influenced certain aspects of the evolution of western culture in general and Canadian culture in particular; but without clear and well-argued criteria for selecting his topics, this only means that Innis had a knack for choosing particularly effective and convincing examples that would capture the interest of his readers. The question, however, is not whether we can entertain our readers with interesting discussions of major technologies that have influenced human development, but whether our choice of one technology over another as an object of study is anything more than strictly random, and if so, what is the principle that ought to guide this choice? It is very hard to find convincing answers to these questions in the writings of Innis or McLuhan.

And this is also the reason why focusing on the title "mass media" does not yet duly narrow down the domain of our research. Allow me to elaborate on this with the help of several examples. Consider, for instance, the story of the discovery and mass production of antibiotics—undoubtedly a fascinating chapter in the annals of how we came to be a modern "mass" (no less so than the oft-repeated story of the invention of the home radio). Or consider the all but mythical story of that modern Midas, the brilliant chemist Fritz Haber. Haber taught the world how to synthesize ammonia gas and so how to produce both nitrate-based explosives and chemical fertilizers on an unprecedented scale. He did this in the early twentieth century, when the traditional means for producing fertilizers (primarily saltpeter) were rapidly running out. Many people are not aware of this, but without Haber, human population would not have reached its current size of seven billion: we simply would not have been able to feed so many mouths. Haber also taught us how to protect the large-scale crops made possible by his chemical fertilizer, by inventing innovative pesticides. Without the chemical fertilizer and the pesticide, the world would not have been able to reach such a troubling scale of overpopulation. The most infamous pesticide invented by Haber, by the way, was Cyclone B, the same gas later used for the mass murder of many of his "fellow" European Jews (Haber was Jewish by descent, but converted to Christianity). Indeed, already during World War I, he was among the pioneers of German chemical warfare. Questions about the wisdom of feeding so many people or the morality of the various inventions introduced by Haber, who was, for better or worse, one of the greatest scientists of his time, are not part of our concern here; my aim is simply to draw attention to the fact that his inventions constitute a central chapter in the story of the processes by which humankind became a "mass" (in the full modern sense of the term, which includes not just our numerical growth as a species but also the mechanization of killing, and the stripping of the human quality from the individual as such).

Trying to move away from Haber's rather horrifying story, I am reminded of the modest discovery by the tragic physician Ignaz Semmelweis. It, too, is relevant to our transformation into a modern mass: Semmelweis taught doctors how to wash their hands and disinfect their tools before attending to women in labor. Many of these doctors had regularly moved between performing autopsies and treating women in the maternity ward without so much as washing their hands with water and soap (as the role of bacteria in causing infections and the importance of disinfection were not yet known). Our becoming a mass owes a lot to such fundamental improvements in sanitation and to the detection, following this new knowledge, of ways to reduce mortality rates. So can we say that the discoverers of Penicillin and chemical fertilizer or the pioneers of modern sanitation are any less responsible for our becoming a mass than Morse and Diesel, Lumière or Steve Jobs? I do not think we can.

Though I imagine that the point is clear by now, I want to mention one more classic example before we return to our investigation. McLuhan was particularly fond of this example and typically considered its investigation no less integral to the study of "Communication" than the writing of the history of the telegraph and its impact on the development of the United States of America. Its origin is in the works of historian Lynn White (1962), and McLuhan cites it with typical pleasure in his famous book about war, one of his best-known collaborations with the brilliant graphic designer Quentin Fiore.[1] I cannot vouch for its historical credibility, but the example is a good illustration of the challenge that we face here. White argues that the stirrup—that little footrest that first allowed the armored horse rider to maneuver on horseback without falling off from the weight of his armor—arrived in Europe from the East around the eighth century. Arrived—and quickly proceeded to change the course of history. The stirrup, he explains, enabled the first effective use in battle of the armored knight, who became more or less invincible as

[1]McLuhan and Fiore (1968), pp. 26–29.

a result. According to White, the cost of maintaining the knight's armor was enormous, indeed unprecedently high for the economy of the time, but the results on the battlefield justified even this enormous expense. A king who wished to keep an invincible army, that is, an army of armored knights, thus had to change the social and economic structure of the population of his subjects from the core: he had to replace the mass mobilization of peasants to fight in the war (the common practice at the time, which produced less than dazzling results on the battlefield) with the mobilization of peasants to provide regular financial support for the hefty expenses of a select group of knights, who were at first just simple able-bodied peasants selected from among their peers and equipped with armor funded by them. Each knight represented in battle the mass of peasants he had spared from the fighting. In other words, the peasant-knight released the masses of his fellow villagers of the burden of participating in war at the cost of enslaving them for the rest of their lives as tenant farmers. The revenue from their labor funded the great expenses of the knight's constant equipment repairs. This is precisely what Charlemagne did, according to White: in order to establish an invincible army, he radically altered Europe's socio-economic structure, rendering it feudal for many centuries to come. And all of this, White contends, because of a small footrest brought over from the East….

At this point, the colleagues whom I inquire about the nature of our field tend to lose their patience. Communication technologies (like the telegraph or radio), they argue, trying one last time to set me on the proper path, are distinct from other technologies that are not communication technologies (like chemical fertilizer or the stirrup) insofar as they are *means of transmitting information*. Pure and simple. The fertilizer and the stirrup, then, are not communication technologies because they are not designed to transmit information. In subsequent chapters I will examine this line of argument closely, for I think that it rests on an error of perspective: in particular, it stems from an essentialistic and even an anthropomorphic misconception. But for the moment, I wish only to point out that it contains significant and fascinating problems that are at the very least worthy of our attention. Not many concepts have produced quite as much intellectual blather as the cluster of concepts we hold about information. As we will see later on, there are quite a few such concepts, and they are not compatible with one another.[2] At the very least, some of these concepts seem to belong to the realm of mathematics and physics, whereas others seem to belong inextricably to the realm of individuals and their environments. Can these realms be united? How? Some experts regard "information" as virtually synonymous with "meaningful content". This view is common among those who regard communication as the transmission of information. For others, however, information is indeed content, but content that has been stripped of many of the features that make it a meaningful entity, with the explicit aim of limiting those features in as much as this is needed, given a certain context or task. However, as we will see later on, in many of these senses, the bacteria that produces penicillin as well as the

[2]For an interesting list of obviously incompatible definitions see Schement (1993), pp. 20–30.

farmer who treats his crops with pesticide or fertilizes them are also conveyers of information, so that it is doubtful whether we could ever, even after elucidating that problematic set of concepts, manage to draw from our basic working definitions any sort of distinction that would be helpful for us all. Samuel Johnson, the greatest lexicographer in our history and a man who thought deeply about the problems of communication and its reduction to information, once noted that his tobacco box is nothing but a physical object until it is sent, empty, to his tobacco seller.... In order to approach this miraculous emergence of an object's meaning carefully, we need much more than an impatient dismissal of our quandaries. Indeed, even if in our attempt to find a satisfying answer to the question 'What is information?" we now went straight to the seminal and most authoritative essay in the history of information theory, by Claude Shannon (the father of mathematical information theory, wrongly called—already by him—"communication theory") and his co-author Warren Weaver, we would immediately come up against a bold assertion that would probably discourage at least some of my colleagues: at the very outset of their essay, *Shannon and Weaver assert that communication is but a general name for any regular influence of one mechanism upon another*. This is almost certainly the source of McLuhan's aforementioned problematic definition, whose loose and open-ended nature caused us difficulty. It is clear, then, that under such a definition, the stirrup, fertilizer, antibiotics, and even hand-washing are all means of communication; they are all media to be reflected upon by the media expert.

When I look out at the classroom, the expressions of my students betray concern. They seem to find my account engaging enough, but the confusion I admit to troubles them. If I go on like this, giving voice to my doubts, I will lose however little of their attention I still hold. Can these students be helped?

Bibliography

Borges, J. L. (1998). *Collected fictions*. New York: Penguin.

Innis, H. (1950). *Empire and communications*. Oxford: Clarendon Press.

Innis, H. (1951). *The bias of communication*. Toronto: University of Toronto Press.

McLuhan, M. (1964). *Understanding media: The extensions of man*. London: Routledge.

McLuhan, M., & Fiore, Q. (1967). *The medium is the massage: An Inventory Of Effects*. New York: Random House.

McLuhan, M., & Fiore, Q. (1968). *War and peace in the global village*. New York: Bantam Books.

Schement, J. R. (1993). Communication and Information. In J. R. Schement & B. D. Ruben (Eds.), *Between communication and information* (Vol. 4, pp. 3–33). New Brunswick: Transaction Publishers.

Shannon, C. E., & Weaver, W. (1949). *The mathematical theory of communication*. Chicago: University of Illinois Press.

White, L. (1962). *Medieval technology and social change*. Oxford: Oxford University Press.

Bibliography

Chapter 3
Appendix from the Classroom: Toward a Useful Introduction to Communication

It is usually around this point in my lectures that one of the more practical-minded students raises a hand and expresses the unease felt by the rest of the class. He means no disrespect, but… putting it gently, he finds protracted theoretical doubts wearing. He wants to find his place in the job market as soon as possible. That's why he's here. He has no time for abstract deliberations. He argues, not without reason, that the professional domains grouped under the broad umbrella of communication departments were assembled somewhat haphazardly, or else for reasons related to financial and social pressures and perhaps not out of any particularly profound theoretical wisdom. His words send a visible ripple of relief throughout the classroom, and I am deeply impressed: he is obviously expressing their unarticulated feelings. In almost every communication class I have ever set foot in, I have found a powerful and surprisingly straightforward desire to abandon the theoretical debate as quickly as possible in favor of matters perceived as more practical: how to master a certain kind of editing software, for example, or how to operate the radio broadcast studio. In my early days as a teacher in the communication department, this trend left a strong impression on me and reinforced my confusion about the seemingly chaotic state of the field's research. I have never found there to be anything less practical than burying your head in the sand without critically examining your goals and your available means for achieving them (recall the fascinating psychology of the overpriced anti-aging cream). Something in the motivational structure of many twenty-first century students seems to be nailed to the hollow capitalist tradition with such force that often they care little whether the theoretical framework that allows them to hold the illusion of "the sure job at the end of the tunnel of school" is meaningful and wise, or meaningless and doomed. Does the practical-minded student not wish truly to improve his understanding of

N. Bar-Am, *In Search of a Simple Introduction to Communication*,
DOI 10.1007/978-3-319-25625-2_3

his environment and of his future?[1] I gently interrupt the flow of my student's animated speech. Give me one more moment, if you will. I do not underestimate the significance of any of the real-life tasks set before you and your classmates. On the contrary: my desire to explore whether and how I can help you achieve these tasks is what brought me to this classroom in the first place. And it is because of this desire that I place these theoretical doubts on our agenda. My aim is to understand if and to what extent you *can* be helped. Can anyone encourage and improve the efficiency of the processes that will allow you to fulfill your goals? Can I do no more than refrain from bothering you with vexing theoretical questions as you study more and more technical details of technologies that will most likely be obsolete by the time you join the workforce? If there is any wisdom behind the grouping together of the communication professions under a single heading it is worth asking what the fundamental tenets of this wisdom are, and teaching them to all of you. This much, I hope no one doubts. If there is such a common wisdom it will unify and explain the practical actions that take place within its framework, and perhaps even enable their improvement. Are there, for example, any principles common to the acts of editing a fine news report and successfully promoting a politician in the digital media? If there is such wisdom, no task is more practical than seeking it. The practical-minded student nods tentatively. Of course, he concurs. But what if there is no such wisdom, he persists, bravely and somewhat anxiously. If there is no such wisdom, I reply, then no task is more practical than acknowledging this fact and adjusting our institutions accordingly. Its absence would imply that there was no point seating students of these various professions in a single classroom to hear a panoramic, strictly theoretical introduction to communication—because such a theoretical panorama simply did not exist. The class would have to be dismissed.

The practical-minded student hesitates. The last thing he wants is to waste his own time, and the prospect of dismissing the class strikes him as rather tempting, as it sometimes does to students who have internalized an ambivalent attitude toward studying; yet somehow, he would still prefer to leave this radical dismantling of academia to future generations. He tries a slightly different tack, which stirs up a certain restless unease in the classroom: the various academic institutions in our field, he says, particularly in the social sciences and humanities, are primarily convenient gateways to various professional possibilities. The top universities

[1]Here I am using an "ideal type" student, of course, so as to deliberately simplify a highly complex and intriguing psychological puzzle that I cannot presently discuss in detail. Let me stress, however, that I have the utmost respect for the realistic urgency of my practical-minded students. It should be noted however that self-deception rarely makes us indifferent to the desirability of learning per se, but rather often expresses our ambivalence toward its hardships. And it exists at *all* levels of learning, from the crudest to the most refined. On the one hand, even people who have been convinced that anti-aging cream is no more effective than water will often continue to deceive themselves about its effects. On the other hand, some forms of self-deception are clearly not just universal but also *biologically beneficial* as means of self-encouragement, of mustering our energies, etc. The slogan "fake it till you make it" has some truth to it, and people who, for whatever reason, feel younger and more beautiful, may on some levels actually become so. Thus, even the mere illusion of becoming experts may on some levels improve our performances.

specialize in this: more than superior knowledge or essential skills, even more than better teachers, they provide a ticket into an otherwise closed club whose select members find in it a life, and in particular a job, that is ready-made. It is genuinely hard to respond to this argument because it combines half-truths with an indirect admission that the question of whether or not we are able to meet the challenge defined here is of no concern to the student. To be sure, I reply, a good part of the social stratification in various countries, in many fields, is upheld by the system of higher education. (Though it is not always true that institutions of higher education aim to train and reinforce the status of an old elite, as is still the case in the United States and Britain; sometimes, like in Israel today, the goal is actually to destroy this elite by voiding the academic degree of its meaning as an indicator of elite status.) It is a fact, for instance, that most of the faculty members who teach workshops in our department are still prominent players in their respective professional markets, and even those who are no longer active professionally still have close contacts with leading figures of the fields they teach. Their acquaintance with the stronger students in the classroom can certainly provide those students with a diving board into the industry. The teacher who is also a newspaper editor might offer his best students the chance to intern at this paper; advertising teachers can open doors at their own firms or those of one of their close colleagues, and so on. But—and this is the important caveat, I think—the whole thing cannot hold together in the long run if it is based on an illusion or a sham. If a program for the general study of communication is no more than a tax and a lip service that needs to be paid on the way to acquiring a profession, then nothing could be simpler than admitting this fact openly and institutionalizing the tax by collecting it directly from new job seekers without wasting our and their time with studies that do nothing to improve their overall professional performance. After all, students are already potential employees; if we can't truly improve their future performance, why not let them join the workforce now and begin without delay to contribute to society and to themselves? Is it only because the system of evaluating students makes it possible, after three years of acquaintance, to make slightly better predictions (and even this is doubtful) about who will succeed more and who less at his or her profession? Moreover, unless we have some sort of theoretical understanding of the requirements of a particular position and of the ability to meet these requirements, how can we presume to predict such success or indeed to grade a student's understanding of the relevant material?

The professed goal of any academic program is first and foremost to raise the awareness of its students to the central theoretical problems of the field for which it offers training, and of course to enhance their ability to solve the practical problems derived from them. The things that enhance one's mathematical wisdom are unlike the things that enhance one's abilities as a physician, and this is why, obviously, the study of mathematics is different from the study of medicine. Thus, it is vitally important not to be tempted by cynicism to shrug off the significant theoretical questions we face here. These questions are: *What is the theoretical knowledge that all communication professionals should possess? Is there any theoretical curriculum that helps transform a layperson into a communication expert? Is there knowledge that enhances our ability to communicate with one another, whatever form our communication takes? Is there any body of knowledge common to all the people who stand out as uniquely effective communication professionals? If so, what is it?*

A student who has been listening to my words with visible tension now raises an interesting point: in every field, there are certain aspects of the skills involved, and of course of excellence, that cannot be learned or taught. The class concurs. I too concur, of course. Who hasn't heard, for instance, of the ice cream vendor who became a billionaire? (You know the type: he would never have made it into any modern business school because he has no high-school diploma or maybe he graduated with poor grades, and anyway he has ADHD, so he would not even have passed the standardized tests required to apply to a prestigious institution.) Few business schools graduates (and fewer of their faculty members) match the ice-cream vendor's drive and entrepreneurial alertness to opportunities for profit. And then—behold—the day comes when the new business school building is named after him. Indeed, it is very hard, and to an extent perhaps impossible, to teach such extraordinary character traits and talents (some of these, like the ice-cream vendor's insatiable desire to maximize his power for its own sake, it is perhaps also immoral to teach even if possible). But we do not presume to teach these qualities here. In no field does education guarantee success, certainly not success on a grand scale: its goal is merely to provide anyone who seeks education with a minimal level of understanding and to give public confirmation of the assimilation of this minimal level. So, for example, economics departments and business schools can certainly offer anyone basic knowledge about the mechanisms of the financial market that will allow him or her to manage their assets and business somewhat more wisely than those who are unfamiliar with these mechanisms; a superficial familiarity with the basic arguments about the proper way to intervene in markets, if at all; and so on. Exactly how this is achieved is a matter of some mystery, not least because most of the market theories that business school students and their friends in the economics departments will meet during their studies are bygone systems describing bygone worlds and dynamics, or ideal systems describing hopelessly ideal worlds and dynamics, and it is not very clear what they can reveal to us about the secrets of the hyper-modern online global economy that shapes our current lives (or how). Perhaps the most popular and widely-studied of these economic theories is that which originated with Adam Smith, the eighteenth-century philosopher and economist. This beautiful theory is founded on the premise that we are all rational individuals who have virtually perfect access to all content that has any possible economic significance. What is it that we can learn from such incredible premises? And on what do business schools teachers base their hope that such theories will help anyone in real life?

It goes without saying that studying at a prestigious business school guarantees neither financial success nor even a livelihood. Education is not a simple mechanical process and its result is never known in advance. But if studying at a business school is nonetheless valuable, and it often is, this is because students at these schools are afforded the chance to be slightly more alert to possibilities of financial success, and so slightly more successful in their financial evaluations and decisions than those who never studied the field. Otherwise, once again, the studies would amount to a pointless tax imposed by our policies upon the final years of youth of our country's future citizens, and society would thus be harming its own chances of growth. And so it is in every field: not everyone who graduated from medical school will become a top surgeon with famous "golden hands", but they will all know certain basic things about anatomy and pathology, and will be able to treat certain diseases with varying degrees of success. Throughout history, excellent physicians have been trained without a formal education: they studied as apprentices. The objective of a formal education in the applied disciplines, then—and this is a crucial point—is not a miraculous promise of high quality and certainly not the promise of a satisfactory job for every seeker but, rather, institutional supervision over a minimal level of service providers. And we have not even mentioned the fine arts, in which excellence is so illusive that nearly no one would presume to teach anything more than its most basic principles. The question, then, is not whether there are illusive dimensions of excellence derived from principles so subtle that they are indiscernible and their uniqueness is inimitable, or else can only be imitated with no literal understanding of their meaning (of course there are such dimensions, in every field: even in the so-called exact sciences, vital knowledge that is tacit—hidden from conscious awareness—passes between researchers, as Michael Polanyi emphasized). The question is rather: what are those dimensions that enhance consciousness and professional skill and which can *explicitly* be taught? What does the minimal theoretical understanding of the field include, and how is this understanding attained? The theoretical programs in schools of communication are justified only if they offer those who pass through their gates knowledge that will ultimately afford them a relative (but certainly not absolute) advantage over those who did not pass through the same gates. Thus the question we are posing here is basic and important: What is this knowledge? What is the basic professional wisdom possessed by a graduate of communication studies and not, say, by an economist or a physician?

These questions invite discussion of several extremely explosive problems, and in the scope of this opening chapter I can only mention a few of these problems briefly—so as to make it harder for us to avoid them later on. First of all, what is normally called the media market—that rather amorphous range of professions whose attractiveness is what draws the bulk of my students to the classroom—is progressing and evolving at unprecedented speed. So rapid is its evolution, indeed, that we may safely predict that *most of the technical training these practical students will receive during the course of their studies will be obsolete by the time they enter the job market or at most several years thereafter*. This, then, is perhaps my ultimate theoretical answer to the practical-minded student: the days are gone when we could make do with practical workshops as tools for professional training in the various fields of communication, since these no longer have the long-term value they possessed before the world began to rush so frantically toward an unknown future. The typical contemporary nation-state is not the ancient Kingdom of Sumer or Babylon, and my students are not the king's scribes. From a social perspective, what is currently happening in our culture is without precedent in human history: the elderly, who had always been the seat of wisdom and experience and therefore the most respected members of every culture and society, now need to be tutored by the young in almost every domain that involves new technology, that is to say, in almost every aspect of their participation in communal life. We are all familiar with this strange scene: young children teaching their parents and their parents' parents how to get by in the world, how to operate a new device or appliance and what contribution it can make to their lives. The young, and not the old, have in many ways become the culture's guides and representatives. Many people find this predicament alarming, and rightfully so, since we all know that the same youth who operate their devices so skillfully still lack any real perspective on the potential influence of all this progress upon their lives. Indeed, no one seems to have such a perspective in our rapidly-evolving world. The interviews I conduct every year with prospective candidates for our department suggest clearly that attaining this perspective is the main motivation for pursuing communication studies, weightier even than the hope of breaking into the job market: our technological environment is changing at a dizzying speed and the most basic need is to achieve some understanding of the dynamic that drives it, of where we are coming from and where we are headed. This frightening dynamic has many challenging aspects. The rate at which technology is currently updated has not quite reached the point that would render professional workshops in communication utterly valueless, but there are already quite a few mass-media technologies in which the students are in many practical senses greater experts than their teachers, who in turn struggle desperately to stay up to date. We will have to devote some thought, later on, to the challenge that such technologies pose to the learning, teaching, and evaluation habits we inherited from our predecessors, since it is clear that the traditional system of education is poised on the brink of collapse due to its incompatibility with current learning environments. Students will draw real value from the workshops only if these practical courses include useful general lessons, theoretical lessons that would enable students to anticipate something about the technologies and interfaces in which they have not yet been trained and whose precise nature no one knows as yet. Otherwise, we have done nothing but train them to tackle a present that is fast becoming the past, and in that case we are no different from the strategy expert who teaches his students how to win the battles of yesterday: we produce experts in wars that will never again be fought as we rush toward a future battle whose nature is unknown. We must therefore investigate the possible objectives of an academic program that unites such rapidly-evolving domains.

Moreover, the constant motion that characterizes the areas of study grouped under the academic heading of communication studies is not coincidental: it is a function of the most salient feature of any phenomenon properly described as an instance of communication. Consider again, if you will, the theory of strategy (warring is undoubtedly a form of communication, however primitive and wretched). And suppose for the moment, even if it sounds ridiculous, that the perfect and complete theory of strategy were attainable. In other words, suppose that we are in possession of a theory that yields the ultimate formula for defeating any army, in the past, present, or future. Suppose we decided to publish this wisdom in a book titled "The Simple Introduction to the Complete Theory of Victory" (such books, as we all know, are almost as common as anti-aging creams). A theory of this sort is supposed to give anyone who reads it rigorous guidance on how to act in every possible war scenario so as to emerge victorious. Now suppose that two generals, commanding more or less equal forces, read the book and set out to fight each other on a battleground that affords neither side a clear advantage. Who will win? I suggest the answer is simple: victory will belong to the general who anticipates the actions of his rival in light of the instructions detailed in the book and then surprises him with new and unexpected strategic moves—in other words, the general who wins will be the one who improves upon or in any case breaks with the rules of the perfect theory…. This paradoxical conclusion is the reason why strategic wisdom will never be expressed in the form of a comprehensive, final theory (a game theory of one sort or another) that provides a fully-detailed and unimprovable set of rules that yields a pre-determined and fixed set of moves. Strategic wisdom involves, among other things, an active alertness to your own ever-changing limitations, to the ever-changing limitations of your enemy, and to the rapidly-changing conditions of your environment. This kind of alertness is antithetical to the fixed algorithm, which dictates actions in advance: the moment our enemies replace their alertness with an inflexible program, they become predictable, and therefore defeatable.

While the example may strike you as too loose and hypothetical to support any general conclusions, and it is certainly true that we will later be called upon to analyze more detailed and practical scenarios, still, I think that it succeeds in capturing something of the fascinating uniqueness of communication as a distinct discipline. This uniqueness is distinctly abstract, but we cannot progress even in our practical-minded educational quests without attempting to understand its roots, since the problems mentioned thus far are all bifurcations of its consequences. Communication is responsible for a fact that I believe is among the most bizarre and fascinating of the many curious and fascinating facts of our world: we are made up of and surrounded by systems that adapt themselves to their environments constantly and in ways that cannot all be accurately predicted by observing the systems' separate components. At least some of these systems modify the sometimes unpredictable rules that govern their behavior as a direct result of their representation and internalization of possibilities for their future behavior in their respective environments. In other words, they study their environments and adapt themselves to them. Indeed, they even shape their environments as they act within them, so that a complex and multilayered reciprocal relationship holds between these systems and their environments. The difficulty involved in making scientific sense of such a reality is formidable, and constitutes the fundamental challenge facing anyone who wishes to study communication. So in the coming chapters I will elaborate as much as possible about it. In a nutshell, the challenge is this: it is not easy to reconcile the fact described above with our scientific picture of the world and in particular with what we know about and expect of a "scientific explanation". The laws of physics cannot be modified or broken. That is why we call them "natural laws" and distinguish them sharply from the laws of the state or the community (described as mere "conventions", for although they are not arbitrary, they are alterable). The Babylonians were already able to predict the timing of the next solar eclipse in their area with impressive proximity, implying that they had identified the existence of a rather rigid cosmic regularity, even if they did not fully understand its underlying causal connections. The natural expectation is that our understanding of this cosmic regularity will not itself have any effect upon the actual course of affairs, since prediction is not supposed to alter the course of the heavenly bodies. Many have noted, however, that it was the prophecy that Oedipus will kill his father and wed his mother that caused his parents to abandon him, and so caused the prophecy to fulfill itself, while many other prophecies clearly lead people to organize effective preventive action, thereby rendering the prophecies false by their very truth.[2] *Unlike cosmology, then, communication theory despite the fact that it describes phenomena that in final analysis undoubtedly belong to the same world described and studied by cosmology, focuses on some regularities that may well be in constant flux, and moreover, on those modes of behavior whose representation, for example through the articulation of a strict regularity that describes the behavior, often leads to a modification of the original phenomena (in other words, it undermines the claim that their earlier representation was indeed an expression of a law of nature).* If, for example, I knew in advance what I was about to write in the next line of this book, I could easily write something else instead.

A Grape-Sugar Spaceship, for Instance.

How can we reconcile this simple everyday observation with our deep conviction that our actions all take place within the physical domain and obey a fixed physical law? This seemingly naïve problem, I suggest, embodies many of the basic difficulties facing those who wish to teach or study communication. It seems, therefore, that we will have to get to know these basic problems as a necessary backdrop for our investigation of more practical aspects of communication.

[2]See Popper (1950a, b) for an elaborated discussion of the "Oedipus effect". See also Popper (1945, p. 22), and the famous introduction to Popper (1957). In the same vein, Robert Merton (1968) famously noted and studied various cases of the self-fulfilling prophecy. A Bank, for example, will go bankrupt for the mere reason that people believe that it will. Both Popper and Merton regarded the "Oedipus effect" as the most distinctive feature of the social sciences. The aim of every "true" prophet, they noted, is to become a "false" one.

Bibliography

Merton, R. (1968). *Social theory and social structure*. New York: Free Press.
Polanyi, M. (1969). *Knowing and being*. Chicago: The University of Chicago Press.
Popper, K. (1945). *The open society and its enemies,* Vol. 2. London: Routledge.
Popper, K. (1950a). Indeterminism in quantum physics and in classical physics: Part I. *British Journal for the Philosophy of Science, 1*(2), 117–133.
Popper, K. (1950b). Indeterminism in quantum physics and in classical physics: Part II. *British Journal for the Philosophy of Science, 1*(3), 173–195.
Popper, K. (1957). *The poverty of historicism*. London: Routledge & Kegan Paul.

Chapter 4
On Questions of the Form "What Is X?", and on the Seemingly Innocent Question "What Is Communication?" in Particular

There is one more introduction I would like to make before we move on (I know: I am trying my readers' patience. But they are autonomous creatures, after all, and can skip this chapter if they so choose). I want to pause here and devote attention briefly to the question I am asked invariably by students, year after year, at the beginning of our second lesson, after the class and I had allowed ourselves in the first lesson to get lost in our fumbled search for the notoriously elusive boundaries of our discipline. When we meet again, my students are visibly tense, and pose a question they believe holds the key to solving their distress: "So what is communication, then?" they ask me, "How do you define it?" My aim in this chapter is to propose a critical examination of the intuitive worldview that breeds this question, and to replace it with a worldview that I believe to be more modest and more sober.

By devoting this chapter to clarifying the philosophical background of my students' question, I hope to illustrate the extent to which all pictures of the world are suffused with philosophical assumptions, including (and especially) the worldviews of those who try to avoid abstract theoretical contemplation or those who mistakenly think themselves to be free of this kind of thought because they are "practical" or "commonsensical" people ("common sense" in many cases designates no more than outdated science and traditions that have become entrenched as a set of default reactions in the everyday conduct of non-critical individuals). In addition, spelling out the assumptions that underlie my students' question will allow me to point out what is, to the best of my understanding, the mistaken common ground of all the prominent models employed in communication research. This last point is an immediately practical aspect of my discussion in this chapter, and it will serve us later on in the book.

Let us look, then, at my students' question "What is communication?", focusing first on its interrogative word. "What is" questions are very common both in the general public and among experts. We ask what is love and what is life, what is the meaning of words and what are numbers. We seek to delineate, distinguish and define almost every topic of conversation, and this mysterious question-phrase—"what is X?"—is an official invitation to erect this kind of boundary. Philosophers have

N. Bar-Am, *In Search of a Simple Introduction to Communication*,
DOI 10.1007/978-3-319-25625-2_4

always maintained tense relations with such questions. Why they are drawn to them seems obvious: the answer to the question "what is X" appears to hold the promise of complete knowledge of the thing we seek to understand, wrapped up in a single statement that describes both the boundaries and the unique properties of the object of investigation. If only we could open every discussion with a few exhaustive basic statements of this kind, it sometimes seems, there would be no real disagreements in our world. This, at least, was the belief of many great thinkers throughout history. Take, for example Gottfried Wilhelm Leibniz, one of the great intellectuals of all time. Leibniz suggested that if we possessed better opening definitions, all major disagreements, both scientific and moral, would boil down to and be resolved by simple pen-and-paper calculations. The kind of statement that supposedly can work such wonders, a statement that encompasses the full extent of a given topic of discussion and enumerates its unique, distinctive features, or properties, is traditionally called an "essential definition". The essential definition is a statement that enumerates all unique properties in the absence of which the topic of discussion will cease to be what it is and become something else. Many of my students dream of organizing everything studied in class into such neat definitions, and they identify the ability to memorize and rattle off these definitions with an understanding of the "material". Despite the fact that this picture of knowledge and understanding is severely distorted, it has a certain tragic practical logic: more and more commonly, exams reflect this misguided conception of knowledge and understanding, thereby affirming and perpetuating the distortion.

The reason for the great caution with which philosophers handle questions of the form "What is X?" is grounded above all else in the simple historical fact that most past attempts to find essential definitions are notorious for their unrealistic pretentiousness. Even attempts carried out in a less pretentious and more down-to-earth spirit, as in some of Plato's early dialogues for instance, are typically known for their lack of success (even when the dialogues end with results that appear to be positive, these results are often unclear with respect to their consequences, and so unhelpful in the eyes of many readers; consequently they have earned the reputation of failures). But it is, of course, not only past failures that account for the cautious approach to such definitions. The caution has to do with the strong suspicion that these failures are not accidental. *Something seems to be amiss with the basic assumption that essential definitions of the sort my students seek are humanly attainable.* If so—and I will soon explain why I think that this is indeed the case—then my students' innocent request that I begin my lecture by providing them with an essential definition of communication, is better left unfulfilled. We will have to find an adequate replacement for this challenge.

To understand the dangers embodied in the tendency to organize knowledge as a collection of essential definitions we need to acquaint ourselves a little more closely with the essentialist worldview. The notion that ideal science is a collection of essential definitions constitutes just one component of an overarching picture of the world that is associated above all with the Greek philosopher Aristotle. The idea that underlies this picture is captivatingly simple: our world, it holds, is a vast storehouse of basic items, clearly distinct from one another; these basic items are what philosophers call "substances". So, for example, according to Aristotle, the person reading this book is a substance, the book itself is also a substance (even if an artificial one), and so is the tree beyond the reader's window. The essentialist picture implies, therefore, that it is possible (at least theoretically) to compile a list of the basic items that make up our world, like a reliable inventory produced by a meticulous, God-like storekeeper: this is the storehouse that holds the universe's natural building blocks. Moreover, according to Aristotle, these natural substances can then be arranged on shelves (that is, into categories and sub-categories) on which they belong naturally, and which reflect unique properties that are necessarily shared by them and only by them. Now this view starts to be rather confusing—despite the fact that it, too, sits well with our common sense (it's startling to notice how quickly an idea that strikes us at first glance as captivatingly simple becomes a convoluted and perplexing picture of the world, which may still seem logical but only because we are so accustomed to it): we do not really know what the words "natural" and "necessarily" designate in the previous sentence, and this fact created serious complications for Aristotle's theory. For as every storekeeper knows, the inventory can be sorted in many different ways, which is to say that many different classification criteria can be applied to what we call, without sufficient knowledge of their nature, "basic substances". Do all sets of items that we can potentially create share an essence? Every beginning student of logic can tell you that there is in principle an infinite number of criteria for classifying the things that surround us, since each item, in principle, has an infinite number of properties. It is only because of our anthropomorphic bias, which leads us to emphasize certain properties over others, that we do not immediately notice this difficulty. Indeed, even if we knew in advance and with certainty that the world is a long list of distinct basic substances in the storehouse of the universe, and even if we knew what these substances are and how to distinguish them from their surroundings, still, from a logical point of view, every substance has infinite properties and each property can be used in order to identify this or that general type to which that particular substance belongs. Aristotle recognized this, of course. He knew that there are many alternative arrangements for every set of substances, and many possible definitions for every set and subset. And yet he nonetheless declared there to be *only one proper, natural, necessary classification of the world's inventory into categories*, classes, types and kinds, and further declared that *this classification is within our reach*, i.e., that all scientists would promptly classify the world according to a unique, complete, and roundly accepted hierarchy. This insistence that carving up the world into sets of substances according to their essence is somehow imminently attainable to the keen observer is the most severe flaw of Aristotle's essentialism, and it is also the problematic assumption that rears its head in my students' question "What is communication?". This is why I find it so hard to answer. The essential definitions we encountered in the previous paragraphs are supposed to be accounts of just such an ideal, unique arrangement: they presume to describe the only natural hierarchy of all the items in the storehouse of the ultimate storekeeper—that is, the unique classification of the world into natural kinds, natural sub-kinds, sub-sub-kinds (no less natural), and so forth, right down to the natural building blocks that are given and indivisible: the truly basic substances.

For the sake of illustration, imagine for a moment that the universe is made up of many Lego parts of various different (but constant) colors and shapes. Essentialist science is the proposition that we sort these parts neatly according to these basic properties so that each unique group of Lego parts has a unique definition, such as "the group of blue parts of size A", alongside "the group of red parts of size A", "the group of blue parts of size B", "…red parts of size B", and so on and so forth until the parts are all neatly arranged. The scientist, according to Aristotle, is she who discovers the essence of each one of these groups, and she is therefore able to deduce from it everything that can possibility be built from each such group of Lego parts.

The intellectual reasons that inclined Aristotle toward the essentialist worldview do not directly concern us here, but for the sake of curious readers I will mention one prominent reason: Aristotle's main area of expertise was biology (he was the greatest biologist of his time, and two thousand years after his death, parts of his system were still being studied with justified admiration). In biology, the classification of the domain of research into animals (which Aristotle regarded as basic substances, as mentioned above) and kinds or species (which Aristotle saw as natural, constant, and necessary) often seems very intuitive (and indeed appears in every known mythology). Groups of animals—such as the group of human beings, the group of horses, etc.—are in certain known senses indeed more natural objects of research than, say, a random group that combines individual items from several of these groups (for example, the group of all human beings from Texas and all horses whose name starts with the letter "C"). From this it is easy to draw the false conclusion that these groups have a fixed and unique nature, an essence that is embedded in the animals themselves and reflects a "deep" and constant structure of existence. It is thus easy to conclude falsely, as Aristotle did and as our forefathers who wrote the bible did, that the species existed always in their present, natural form and will remain this way forever (or in any case so long as a certain species does not become entirely extinct). Consider, for example, Noah's ark: one by one, the long queue of animals enter, featuring at least one male-female couple of all present species (we will set aside the prosaic fact that not even an aircraft carrier could contain such an astoundingly varied group, as well as the fact that not all animals reproduce by mating). It seems that the ideal to which Aristotle aspired was effectively to peer over Noah's shoulder, note the name of each eternal species entering the ark, and attach to it its proper and unique Lego-part-definition, which distinguishes it from all the many other queuing animal species. As Aristotle pointed out repeatedly to his students (and with good reason): human beings beget human beings, not horses. This observation can appear to justify essentialism, but this is a false impression, since the justification is circular: from an evolutionary perspective, there is no doubt that what we now call "Horse" was born of an animal that in the distant past was very different.[1] What appears like an eternally fixed species is often just an approximated abstraction of constant change occurring in the details we observe before our eyes, and this approximation, or idealization, always expresses a conjecture about which details are important—that is, important relative merely to the interest of a given observer at a particular moment and no more. It reflects at most a certain set of aspects of reality on which we have chosen to focus.

But how are these abstract ideas related to the world of communication theory? It is a fact that underlying the essentialist picture of the world is an *illusion of constancy* caused by a basic confusion between the status of ordinary-language categories (the categories of common sense and the average dictionary) and the status of the categories that deserve, perhaps, to form the basis of the perfect natural science at the end of time (whatever may be the meaning of this vague term). The essentialist illusion is grounded in our sense that we presently already possess a perfect grasp of the fundamental and unchanging Lego parts that constitute the world. *Because this grasp appears to us complete, and because we feel that it expresses the ultimate truth about the world, it does not strike us as context-dependent*. This is why it gives rise to the rather incredible assumption that a teacher of communication can (and should) open his lecture with a perfect definition of his field. And what is more: *it gives rise to the truly incredible view that there are no real problems of communication in our world*! Since, if the world is an inventory of fixed and widely known substances, if there is no real disagreement over their identity and proper classification, then clearly each of these substances can be given a name, a marking label. And since the substances are all arranged in their uniquely proper arrangement on "natural" shelves (which, being natural, are themselves similarly free of any context and observer), these shelves can also be given names. It follows that the various human languages are little more than collections of different sounds attached to precisely the same basic ideas and the same basic combinations of ideas, which designate precisely the same basic substances. *According to the essentialist worldview, then, when human beings use*

[1]Darwin, of course, also classified animals into different groups and species. But in his account (as in those of his prominent predecessors, like Lamarck), the species "undergo" constant change (slow, but constant), and therefore have no essence at all. In Darwin's account, all the species are descendent from a single ancient origin (or at most from a small number of sources). A species for him is nothing more than a collection of individuals capable of fertilizing each other and producing fertile offspring. This observation appears to echo Aristotle's statement about horses and people, but in fact betrays its arbitrary aspect in a way that cannot be avoided, since the ability for mutual fertilization is itself a property in constant flux and is often the result of markedly arbitrary circumstances. If we were being strict Darwinians, for example, we would be forced to admit that sterile mares are not part of the species Horse (and sterile women are not part of the Human species). Or consider African elephants. They cannot mate with Asian elephants just because of the vast distance between the two groups and not due to a physiological limitation of any sort. These two groups cannot produce fertile offspring, then, for reasons that are wholly accidental. But for Darwin, this is a sufficient reason to regard them as distinct species. According to him, they may become a single species if and when they are brought together artificially, for instance in a zoo, so long as they are able to produce fertile offspring together.

language they are all talking about the very same Lego parts, and when they become scientists they learn to classify them in exactly the same ways (the same common nouns, verbs, etc.*). In other words, we all speak basically the same language, and moreover, it is a perfectly accurate language, as if every Lego part in the world carried its divine name on its back.*[2]

And so, if we come across another person whom we do not understand, it seems that the only explanation is that she is using different sounds to designate the same things. If we want to understand her, at most we need to learn a new set of sounds that designates the same Lego parts that everyone is talking about. We all share the same glossary and need only to agree with others on joint labels for these terms: from then on, we will have no more problems of communication. (The only exception to this rule arises out of certain secondary technical complications that we will discuss later on; for instance, cases in which our interlocutor's words are momentarily blotted out by noise.) In short: behind my students' innocent question lurks the essentialist picture of the world in which is ingrained the deeply mistaken belief that all languages are based on one ideal, archetypical language because the world itself is an unambiguous, vast but perhaps not overly complex Lego structure. Behind their question lurks the false belief that a perfect understanding of the other, any other, is built into our world and into our "selves", and furthermore that real scientific disagreements cannot exist. I am convinced that nothing is more important and more practical as a first step toward studying communication than ridding ourselves of this false picture. When successful communication occurs between people, it is a wonder to behold, and the marvel should not be minimized by the fact that it is common. And we must not eliminate the wonder by giving it simplistic explanations.

It will probably come as no surprise to learn that the few spot-on accurate definitions found in Aristotle's writings are of mathematical objects and of artificial objects, such as tools. Time and again, he returns to these examples when various problems arise in his futile search for truly essential definitions of the things around us. This fact is unsurprising because both mathematical objects and physical tools tell us very little about what are supposed to be the "basic substances" that make up reality. Mathematics establishes an ideal world through mathematical definitions, or axioms: this world is aligned perfectly with the definitions precisely because they are axioms: they cannot be wrong as a matter of principle. (If, for example, we replaced the definition of a triangle with that of a circle, from that moment on we would be investigating the properties of the circle under the name "triangle", but the definition itself would not thereby be falsified). At most, we can produce definitions that are uninteresting form the point of view of the mathematical community.

---

[2]This false conception of language is often described in the literature as the "Augustinian" picture of language, after a ridiculing description of it that appears in Wittgenstein's **Philosophical Investigations** (where it is used in effect to ridicule a view expressed in Wittgenstein's own early philosophical work, **Tractatus Logico-Philosophicus**). Incidentally, in the Jewish tradition this same false picture rears its head in the beautiful creation myth that tells of Adam, the first human, giving every animal its true name.

(If, for example, we redefined a circle as "that which is common to all circles and to the right-angled triangle", we would simply find ourselves investigating objects that mathematicians would judge as uninteresting, but we would not be making a mistake in the description of the structure of reality.)

The definition of artificial objects and of tools is similarly not a typical case of the essential definitions that Aristotle required the scientist to produce, since the tool, as he himself emphasized, becomes a tool by virtue of the intention of its user while using it. A piece of metal and plastic, for example, becomes a screwdriver because that is the use someone makes of it or assigns it at a given moment. If we use it instead as an ice pick or a back scratcher, it will become, by virtue of our intention, an ice pick or a back scratcher. This is not even a matter of how well the tool suits the role that we have assigned it: a lousy screwdriver is still a screwdriver and will remain a screwdriver so long as we insist on using it to turn screws. Its "essence" therefore (insofar as it can still be described as such, since this "essence" no longer designates anything necessary, natural, context-free or eternally constant), will change according to and by virtue of the intention of the screwdriver's user and the context in which it is used; and this, of course, is at odds with the definition of an essence as something objective and inherent, that is to say, embodied in the object by virtue of its nature and independent of the observer and the context of discussion.

Common to the pure mathematical definition and the functional definition of the tool is the fact that both are verbal, or nominal, definitions. Nominal definitions are true only in an empty sense: with these definitions, we establish a dictionary within whose bounds we then embark upon the long search for the truth about the world around us. (From time to time, we even update the dictionary itself in light of our evolving knowledge: concepts such as "ether" drop away from the dictionary, while others, like the word "melancholy"—literally: black bile—remain but lose their original meaning. But let us not go into that now.) When we leave the realm of nominal definitions and turn our gaze upon the vast and mysterious world, immediately we find ourselves facing the inescapable human condition: namely, our profound ignorance regarding the ultimate structure of the universe (even the very existence of such an ultimate, essential, structure is doubted by some scholars, who believe that the very attribution of a particular structure depends on the contingent emphasis and interests of an observer). Clearly, no one has introduced us to the universe's basic "Lego parts" and we would all be wise to adopt a measured modesty about our ability to achieve a firm grasp of such parts in the near future. And yet, communication is taking place between us at this very moment! What is it, then, that makes this possible?

I want to emphasize that I am hereby taking no stand whatsoever on the ultimate veracity of realism or its relation to essentialism. There is no need to do so: even if the universe is indeed made up of fundamental basic items, the crucial fact remains that we are *still searching* for their identity and nature; the items that currently underlie our picture of the world are therefore most likely *not* the basic items that make up the world. Thus, *since these items are at most approximations of the truly basic substances (whatever that means), they lack any real essence in and of themselves*. It is true that the electron and the quark are examples of units that we do not currently know how to break down any further (notwithstanding various theories that speculate about the ways in which this may be doable in principle), and therefore we regard them as constituting fundamental units of existence in a certain known sense (though they are far removed from the world of Aristotle's substances, for reasons that I will not go into here)[3]: to date, they are indeed modern physics' basic Lego parts of a certain sort. But this does not mean that they will necessarily also be the building blocks of any future science, and we certainly know of no real and interesting connection between their being the currently assumed building blocks of the universe and the fact that you and I are currently succeeding in communicating as writer and readers. No scientist holds a pre-given list of the world's fundamental items, and even the question of whether any such constant items exist cannot be determined without research and investigation. This basic point was nicely underscored by Bertrand Russell when he mocked the staggering pretension of Aristotle's contention that here he already holds the list of essences of the list of substances that the scientist will discover, perhaps, in the far distant future. Russell was amused, for example, by the fact that traditional lists of basic material elements (earth, fire, air, water, metal, etc.), perceived in various philosophical traditions as a list of (basic) substances, includes such items as wood and air, which we now know to be compounds of more basic "elements" (which are themselves not elementary particles of the known reality). Thus they are not fundamental substances at all, even in the simple and more lenient chemical sense of the term. Other basic elements on the list are closer to what we now describe as processes, like fire, which is a relatively rapid oxidation process and therefore does not meet the traditional criteria of a fixed "substance" at all. And despite all of this, people understood each other's meaning fairly well even in those "dark" days.... Modern physics, as Russell already noted, altogether abandoned traditional substances, and thereby rejected the essentialist picture of the world that lives on in the common sense of many practically-minded people, the picture that regards our environment as a fixed setting (empty space) containing discrete items that persist in their unchanging nature, like the child's Lego parts or household furniture (Russell 1946 p. 207). The discrete items of the ancient world are the atoms, of course (*atomos* literally means "indivisible" in Greek, and its Latin translation is "individual"). And here we see that the original atoms are ones that we have to a large extent already succeeded in dividing further, and that of the empty space said to contain them—nothing remains. Instead of this picture, as Russell pointed out, we nowadays take the universe to be a non-empty, warped continuum of space-time, we take "matter" to be based in energy, and various other "oddities" that do not fit easily with the traditional essentialist picture of the world.

---

[3]The talk of "substances" has of course become largely dated and not exactly "kosher" in its application to the elementary particles of modern physics. This is why I prefer the term "basic items" in the chapters to come. Interested readers may want to consult the classic but still very helpful Schrödinger (1957) , 192–223.

In sum, the essentialist picture of the world is based on an illusion and encourages it. The illusion is that our ultimate wish is already granted, that we already possess a solid grasp of the unchanging, context-free, basic building blocks that make up the world and by virtue of which we can communicate successfully. This illusion is rather childish, and in Aristotle's case also suspiciously anthropomorphic (Aristotle tried, for instance, to define the essence of poison, whereas it is clear that no substance is essentially poisonous). The illusion is nourished by common sense, which reflects at most the temporary conventions of communication among the members of a given community and not a profound truth about the ultimate nature of the universe, nor even about the binding conventions of other communities. More troubling still: the essentialist picture of the world justifies itself by virtue of a kind of feigned innocence on the part of common-sense people, who proceed as if our conventions of communication already express a set of obvious truths about the fundamental parts of reality. But once we try to examine its stability, the illusion collapses immediately into the temporary dictionary of the nouns of our language, on the one hand, and the unknown universe whose nature we need to explore to the best of our ability, on the other.

Once this crucial doubt is introduced and we realize that common sense does not possess a true grasp of the fundamental Lego parts that constitute the world, the question of our successful communication emerges as an unparalleled mysterious wonder deserving of profound investigation. Indeed, even the question of the possibility of studying this wonder requires extremely careful thought: if we are not in direct contact with eternal basic substances that dictate the dictionary of our language, how is it that we are able to understand one another? How does our communication, which is sometimes successful, occur? How have we arrived at an agreement on the meanings of our words—that is to say, how have we come to be a community of speakers? How do we initiate our children into our verbal agreements? And perhaps most importantly: why and by virtue of what do words possess meaning?

The central weakness of the traditional essentialist worldview was nicely summarized by Karl Popper[4]: Traditional essentialism, he explained, instructs us to begin our scientific investigation only *after* this investigation has in effect been completed. It invites us as a first step to list the essential definitions that we may perhaps possess in the far distant future. In other words, it tempts us to pretend that our ultimate wish is already granted. If we look at the various sciences, even the most exact and rigorous among them, and if we can agree that they have made at least some progress since Aristotle, we will realize that Popper was right: biology advances at a dizzying rate, and despite various faltering attempts, no biologist is currently able to open his Intro to Biology lecture with a decisive, final definition of the living organism. Psychology also makes frequent progress, yet no one knows what consciousness is (or, if you will, what the soul is). Even mathematicians, whose object of investigation, as already noted, is an ideal world bound and created

[4]Popper (1945), Chap. 11, especially Sect. 2, pp. 9–21.

by definitions, smile sheepishly when they are asked about the eternal boundaries of their field of study (which started out, in a certain known sense, as the study of fair trading transactions and the study of size ratios between agricultural plots, and expanded from there into entirely unexpected domains). It is not surprising, therefore, that communication scholars are also unable to open their introductory lectures with a perfect definition of their discipline. That is how things ought to be, at least until we gain divine wisdom.

There, I tell my students: we have just succeeded in riding out a very theoretical discussion. Its immediate conclusion, however, is thoroughly practical: we need to reformulate the basic challenge of the introduction to communication.

But if we cannot begin an investigation with a perfect definition (that is, with the end-result of the investigation), my students ask me with concern, if we cannot open an investigation with a perfect demarcation of the object of our study from a God-like perspective, how then will we know what we are studying? Communication, after all, is nothing like a common-sense material object (even if imaginary), such as a screwdriver or an ant: you cannot grasp it in your hands. How do students and scholars of communication, and especially those who wish to study communication-in-all-of-its-forms, know that they are facing a phenomenon that belongs to their discipline?

This is the question I will try to answer, insofar as it is in my power, in the next part of the book. But a few comments are in order here. First, we typically have no difficulty identifying distinct cases of communication, even without any adequate definitions. The same is true of most domains: definitions are not tools for identifying objects but at most a way of unifying conjectures about them. Second, the rejection of essentialism is not at all a rejection of the attempt to find accurate and useful working definitions, rather it is a rejection of the illusion that our working definitions offer an exhaustive summary of our future, final scientific findings, that they already contain, on a small scale, everything that we can ever know, like the axioms of the deductive system necessarily contain everything that is derivable from them. The rejection, then, is of the status of our working definitions as eternal axioms. So before turning to look for distinctive characteristics of communication, we should ask how researchers in other disciplines handle the lack of essential definitions. How, for example, in the absence of a perfect essential definition of their field, do mathematician know that they are facing a mathematical phenomenon? The answer to this question is fairly simple, and has two parts: first, contrary to what is sometimes taught in overblown and dubious introductory courses, especially within the social sciences, most honest and sober experts admit wholeheartedly that they do not know the precise boundaries of their discipline. The history of mathematics is full of fascinating examples of these unclear, shifting boundaries: an original mathematician invents an innovative method of proof or a new mathematical object, and the mathematical community, or part of it, is initially very skeptical or even dismissive. There is a delicate and open negotiation of these questions in every science. Second, and no less importantly, the word "discipline" itself contains problematic traces of the essentialist worldview and only once we rid ourselves of them will we be able to use it in a reasonable way: the traditional,

Aristotelian discipline was supposed to have been defined and delineated in advance and eternally by essential definitions, that is, by a claim to circumscribe the talk about a certain naturally distinct set of objects. Yet even mathematics is not a discipline according to this definition, since we have just noted that while it deals with ideal, eternal, objects, its precise boundary is in flux and even a matter of subtle negotiation. Despite all the protestations of our common sense and of Aristotle, then, there are in science no truly eternal disciplines. Of all the scientists, mathematicians are perhaps the closest to being experts in a veritable "discipline", almost in the traditional sense of that word. This is because formal mathematical definitions, as we already noted, largely (but not always, and not entirely) constitute the mathematical world of objects. Mathematicians, however, are too busy doing mathematics to concern themselves with questions about the demarcation of their field. To a large extent, the word "discipline" designates for them simply a certain sum, but not one that is set in stone, of central problems, and a collection, also not set in stone, of tools for solving them. We would do well to follow their example and draw from it a general lesson: *we only truly rid ourselves of the essentialist worldview when we replace the notion that a discipline is determined in advance by basic definitions with the notion that a discipline is a delicate negotiation of basic problems*.

(Of course, that the discipline is not set in stone and the tools are not a predefined, sealed set does not mean that "anything goes" and every ignoramus is immediately accepted into the community of researchers as an expert. Any mathematician who cannot show how the new tool he invented helps his colleagues solve common problems that concern them as well will not have their ear.) This, then, is the sense in which we will henceforth use the word "discipline": a collection of common problems studied with the tools accepted by members of the community that is interested in those problems. Communication is no exception to this rule. If we want to hold a general and introductory discussion of communication, we need to start not with a preliminary definition of the discipline's essential boundaries but by proposing a general problem that is common to the researchers whom we are addressing. Thus, *our first step in the search for the simple introduction to communication will be simply to try to articulate such a general problem. Then we will need to examine how we should approach it, and how the potential solutions to this problem become practical tools in the hands of the communication expert*, and even in the hands of the "practical" student, the one whose interest in abstract theories is limited. This is the task set before us here.

Bibliography

Agassi, J. (1997). Who needs aristotle. In D. Ginev & R. S. Cohen (Eds.), *Issues and images in the philosophy of science* (pp. 1–11). Dordrecht: Kluwer.

Bar-Am, N. (2008). *Extensionalism: The revolution in logic*. Dordrecht: Springer.

Bar-Am, N. (2012). Extensionalism in context. *Philosophy of the Social Sciences, 42*(4), 543–560.

McKeon, R. (Ed.). (1941). *The basic works of aristotle*. New York: Random House.
Popper, K. (1945). *The open society and its enemies* (vol. 2). London: Routledge.
Russell, B. (1946). *A history of western philosophy*. London: George Allen and Unwin.
Schrödinger, E. (1957). *Science, theory and man*. New York: Dover.

Part II
Toward a Philosophy of Communication

Chapter 5
Emergence and Reduction

I do not know to what extent this book is philosophical. I therefore do not know how it will serve readers from the field of philosophy. But I do hope at least to pique the interest of its readers in the philosophical problems that underlie communication studies, problems that I now aim to present. This part of the book is devoted to problems that ought to concern students of communication and students of philosophy alike. I found no way around presenting them at this point, even if this makes the following chapters rather abstract and therefore not an easy read for some. The reason is simple: the theoretical problems that I will now present translate quickly into practical problems that students of communication cannot avoid, and into methodological questions with which researchers and teachers of communication must become familiar. I think the authors of introductions to communication who have come before me paid a high price for attempting, often in rather clever ways, to avoid a discussion of these abstract problems. At stake here is the very possibility of studying communication, the dependency of communication studies upon neighboring disciplines (such as biology, psychology and sociology), and the prospect, given this multilayered dependency, of systematically studying and improving communication-related skills. I think that once we turn our attention to tackling these problems (and in particular to considering the basic premises required in order to allow for the proper study of communication), our theoretical world will be thoroughly transformed. And I am not referring only to the theoretical world of communication scholars. I am talking about a fundamental change in the scientific worldview as a whole. In my view, the very recognition of the existence of communication reveals deep-seated limitations of the commonly accepted scientific worldview, thereby provoking us to reevaluate our traditional set of fundamental problems of knowledge and our usual modes of handling them. I suggest, and I will explain this suggestion in some detail toward the end of this part of the book, that *those who take the existence of communication as their starting point will quickly find themselves regarding the main traditional problems of philosophy as uninteresting at best, perhaps even as pseudo-problems. And they will regard the usual methods of tackling them as non-practicable. They will then feel the need to replace these traditional problems with more interesting and meaningful ones and to approach them with more practicable methods*. (Who knows, maybe this will even help to rehabilitate the poor status of philosophy within the general public.)

N. Bar-Am, *In Search of a Simple Introduction to Communication*,
DOI 10.1007/978-3-319-25625-2_5

Our work plan is as follows: I open with a brief and rather general presentation of one of the broadest and most basic controversies that I am aware of—the debate between reductionists and emergentists. The reason for presenting this debate here is to allow me in the next chapter to articulate a closely related problem, one that I consider to be the fundamental theoretical problem of communication studies, the challenge that all students of communication share (or should share). The problem is this: How to unite the most puzzling cases of emergence that we know under a single theoretical umbrella? The partial answers I offer here will be put to the test in the book's final part, where I will explore their usefulness as tools for advancing the skills and expertise of scholars (or researchers) and practitioners (or professionals) in the field of communication. I will consider, first, cases and dilemmas that have preoccupied researchers of communication in the broadest sense of this term, and then issues that concern researchers of communication in the narrow sense, encompassing only those disciplines that branch out clearly from the social sciences. I will of course also discuss the various limits that the general problem I articulated imposes on the scientific status of research in the field of communication in particular and in the social sciences in general. And that's about all. Our introductory journey will thereby reach its endpoint.

A quick reminder: in previous chapters, we noted that "communication" (in the most general sense of the word) is a term that unites aspects drawn from almost every field of human research. This is why it attracts researchers from various fields who seek a more general overview than the one afforded them by their particular area of expertise, including, in some cases, overeager and hasty researchers looking for a shortcut to a general theory of everything. Even highly professional and responsible researchers who established interdisciplinary research groups that brought together the top minds from the world's top research institutions with the aim of standardizing their joint interest in communication and establishing it as a bona fide scientific discipline quickly fell between the institutional cracks. Many of them were forgotten because they were not associated with any well-defined academic tradition, or else are known only within a certain professional circle thanks to some sort of technical contribution they have made to the world. I mention this dynamic to draw our awareness to the fact that those who seek out general guidelines for communication studies tread a fine line: communication, as I will argue below, and in any case communication-in-the-broad-sense-of-the-word, which is the object of our attention in this part of the book, is an inter-inter-disciplinary field: it brings together selected aspects of many of the prominent interdisciplinary fields of our time, from sociobiology to the science of networks. The main questions that we will therefore need to address concern the possibility of conducting research at such a general level, the reasons for doing so, and the right way of going about it. (This does not mean that every person working in the field of communication must consciously know all of the details of this study, but it does mean that anyone who wants to guarantee a minimal level of understanding among newcomers to the field must develop their awareness to the issues raised here—and that is our present goal.)

The interdisciplinary nature of so much of what has been going on in recent years in higher education is not a passing trend. It appeared on the academic scene as an implicit admission that the boundaries imposed on us by the traditional Aristotelian notion of a discipline, or field of study, are simply too rigid. Definitions cannot isolate for us the subject matter of a science and distinguish it perfectly and permanently from the subject matter of other sciences, because science is a quest not a catalogue of results. In particular, certain central difficulties that preoccupy scholars of traditional disciplines appear to be unsolvable within the confines of these disciplines, their solution requiring the cooperation of other researchers from what the traditional system would call neighboring fields. And this, in turn, raises the question of how this cooperation ought to be conducted.

To bring us closer to the heart of the matter it would be best to begin the discussion using terminology that is as intuitive and straightforward as possible, and in subsequent chapters to adjust and refine it gradually, as the need arises. First, then, we need to note a very basic fact about the predominant scientific worldview: it is distinctly reductionist. Reduction is the most basic explanatory tool of science, and many hold it to be the only legitimate tool. The scientist uses it to remove a problem from one field by reformulating it in another field, considered more fundamental, so as to solve it "over there". In other words, reduction is an act of interdisciplinary translation, carried out according to meticulous rules of deduction: whenever reductionists come across a term from one traditional discipline (say, "water" in the descriptions of biologists or ecologists), they replace it with a term from a discipline that they regard as more fundamental (for example, chemistry's "H_2O"). Such a translation is supposed to be exhaustive in the sense that all the qualities of water known to the biologist and the ecologist (that it is essential to life, clear, expands in volume when freezes, etc.) are fully explainable as consequences of their chemical makeup. In this way, reductionists hope one day to do away with all the scientific theories but for one basic theory of the most fundamental elements that our world comprises. They hope in the final analysis to get rid of all sciences apart from physics (and not even all the branches of physics but only the most elemental branch, which will deal with the simplest known particles).

The reductionist worldview, then, expresses the hope that one day (however far and ideal) all solvable scientific problems will be solved within particle physics, i.e., that all scientific problems will be translated into problems of physics and solved as such. While it is true that areas of research far removed from physics, such as the social sciences and in particular a field like the one that concerns us here —communication—appear to us today to exist in a parallel universe (and for this reason there are those who even scoff at the branding of these fields as "science"), the hope, nonetheless, is that this separation is a temporary matter, at least with respect to those problems from within these fields that lend themselves, in principle, to satisfactory scientific solution. Crudely speaking, we may say that the overall aim of the reductionist is to reduce all social sciences to psychology, psychology to biology, and biology to chemistry and physics.

The common reasoning behind this hope goes more or less like this: today there is hardly an intelligent person who does not recognize the fact that everything in our world exists in the space-time continuum. And here we have physics, which is the study of space-time (more recently, it is also the study of the coming into being of space-time from more basic elements, but we won't get into that). It follows, that if a certain science—biology, psychology, or sociology—studies aspects of reality that are not ultimately translatable into parts of physics, then there would appear to be aspects of reality that are not ultimately parts of space-time. And this sounds absurd. At least it does to the ears of the majority of the scientific community. Thus, reductionism as discussed here is simply an expression of the view that the most complete physical theory that we will ever have should also be our most complete theory of everything we see before us and can at all be explained through science.

And yet, strange as it may seem to those who encounter such claims for the first time, a growing number of scholars and scientists contend that the reduction of all scientifically-solvable problems to the science of physics is an unattainable fantasy and will remain so even if we imagine ourselves to be standing at the end of all days in the temple of God basking in the light and glory of His complete physical plan. In other words, many thinkers (and their ranks are swelling steadily) believe that even if nothing exists nor can exist outside space-time, physics will never become the complete, unifying theory that explains all the fascinating and scientifically-explainable phenomena known to man. For example, they argue, physics cannot suitably replace the study of human behavior, certainly not in a manner so complete as to render redundant the descriptions of this behavior in biological, psychological, sociological and economic terms, and so, also, to eliminate the interdisciplinary field known as behavioral science (which is of course an important component of the study of communication in the broad sense of the word). According to these thinkers, then, central aspects of the behavioral sciences cannot form an integral part of a more "basic" science no matter how developed, detailed and complete this basic science comes to be.

Their argument is that our environment is inundated with phenomena whose reduction and explanation in terms of the rules and mechanisms of the "basic" sciences seems fundamentally misguided: it fails to capture the nature of the phenomena that we set out to explain, or at the very least restricts our explanation in ways that we cannot afford. This argument may at first seem vexing, for who can honestly presume to predict our knowledge in the distant future? Yet it is important that we get to know these phenomena now, since the main observation that I will offer in the next chapter is that the question of how we may nonetheless go about studying the most fascinating of these phenomena is the basic problem shared (consciously or otherwise) by all those who study communication.

As a first step, let us take on board without argument a definition that is very common in the scholarship; later, we will refine and elaborate it in various ways. For now, let us describe as "emergent" any property of a system that, to the best of our knowledge, cannot be attributed to the system's components individually. An elementary particle, for example, can have neither color nor temperature, and a single molecule of H_2O cannot be wet. Emergence as we have just defined it is not a particularly surprising phenomenon, and this is a fact that I want to stress because many people are confused about it: emergence is the natural state of things in our environment, because it is invariably true that even a given set (let alone a system) will have certain properties as a set that none of its members can possess individually. Take my wife and me, for example: we are a couple, even arithmetically, and obviously, each one of us individually is not a couple. But it is important to note also that it is relatively easy to explain why one plus one makes a couple (at least arithmetically this is simple; the sociological phenomenon of couplehood is rather mysterious from the point of view of physics and arithmetic). A four-pound pile of tomatoes is also a simple sum of the weight of individual tomatoes in that pile and it would indeed be odd if each of them separately also weighed four pounds. In the same way, even the property described in ordinary language as the "wetness" of water, for years considered an almost paradoxical phenomenon (water is hydrogen and oxygen: the former is highly combustible, the latter is what allows combustion, and yet water puts out fire!), is now explained by reference to several of the statistical properties of the water molecule (H_2O).[1]

However, and here we arrive at the heart of the matter, there are also certain emergent properties that can and should be described as "puzzling", and they are the ones that will interest us here. An emergent property of a system is puzzling when it not only does not belong to the system's constituent components individually but is also such that we know of no way to explain its existence as resulting from the mechanisms that describe the behavior of the system's individual

[1]Statistical explanation limits reductionists in a number of ways that we will not dwell on here. Let me just note briefly that when we attribute statistical properties to entities that we regard as fundamental, we commit ourselves to a position that has significant limitations from an explanatory point of view. Consider, for example, the correlation between smoking and deaths from lung cancer. We no longer doubt that smoking raises the risk of the disease as well as the risk of dying of it. But since this is just a correlation, since some smokers will not get lung cancer, we cannot treat smoking as a satisfactory explanation of lung cancer among smokers, and even less so as an explanation of a particular smoker's illness, because again, his neighbor who smoked just as heavily did not get cancer. Traditional scientific explanation is deductive: it guarantees a certain conclusion given certain premises. The illness of the sick individual does not and cannot follow from the premise about the correlation with smoking because his neighbor's non-illness is also compatible with that same premise. We may, of course, go on to search for other causes for the cancer, but that would just be to deny our basic assumption that the correlation that we have found is fundamental, that is, that it expresses our best possible knowledge. Alas, whenever we attribute statistical properties to fundamental entities, that is precisely the case. This is why statistical explanations greatly limit the reductionist.

Returning to the reduction of the ecologist's "water" to the chemist's "H_2O", it is vital to note how quickly common sense forgets that this reduction is an idealization—that is, ultimately no more than a correlative *approximation*. It is a fact, for instance, that if humans and other animals drank H_2O they would be dead within hours—from a drop in vital minerals. If the water we drink did not also contain these minerals in addition to H_2O, we would not regard it as a source of life. This is why a hasty reduction of the ecologist's "water" to the chemist's "H_2O" does not allow us to provide a satisfactory explanation of many phenomena known to the ecologist and the biologist. *All idealizations, then, leave some aspects of reality unexplainable in principle*. Indeed, almost all the properties we regularly attribute to water are emergent in this sense: water expands when it freezes but a single H_2O molecule does not, water dilutes other substances but the single water molecule does not, etc.

components: even if we imagine a perfect theory (as perfect as is humanly attainable) that gives a full description of the behavior of the system's components individually (as full as is humanly possible), we still do not see how these rules can in any way account for certain aspects of the behavior of the system as a whole. If we have identified such systems in the world, then in them we have found an aspect of the known reality that we do not currently know how to reduce successfully.

The most common cluster of examples of puzzling emergence offered in the literature involves those properties that for generations have led human beings to regard themselves as possessing a soul (and to regard the soul as existing in some sense beyond the material realm—that is, beyond physics, beyond space-time). For example, the fact that we are capable of subjective feeling, of experience and emotion, and even in some circumstances of original thought and creative action. Even modern psychologists, who deny the existence of an eternal non-material soul, clearly still describe us and regard us as feeling creatures who possess self-awareness, a will, learning abilities, an ability (albeit limited) for rational planning and an ability to carry out these plans—in other words, they treat us as creatures who are (to a limited but significant extent) free in their thoughts and actions. We cannot violate the physical laws of nature, obviously, but this does not mean that these laws, even if known to us, would allow us to explain everything that is in principle explainable in our world. Emergentists seek first of all to draw our attention to this fact. They point out that, as far as we know, the stuff that we are made of—the atoms studied by the physicist and even the cells of the nervous system studied by the neurobiologist—does not possess any of the surprising properties that our consciousness displays, and therefore we do not know how it is that they emerge from the properties we attribute to our material constituents: the single neuron, for example, shows not even the slightest trace of an ability for abstract thought or self-awareness, to say nothing of free will or freedom of action. And what little knowledge we currently have about the ways in which neural networks grow and connect brings us no closer to understanding how such surprising properties emerge. How, then, emergentists ask, are we supposed to be able someday to reduce these properties to a theory that deals exclusively with the familiar mechanisms of neural connections, or indeed with the familiar mechanisms for the bonding of atoms in all their various properties (statistical and other)? If we try to do this, it seems that we will find ourselves eliminating the distinct uniqueness of the phenomenon that we set out to explain.

It is important in this context to recall that the problem of the emergence of consciousness from matter has its roots not only in the foundations of scientific methodology, which enjoins us to break down phenomena to their elements in the hope of understanding how they occur, but also in the basic metaphysics on which the modern scientific revolution was founded. Since the days of Galileo Galilei and René Descartes, our world is split sharply in two. On the one side, we have the reality known as "external" and described as comprising bodies (amenable to geometrical analysis) and forces (which can be measured and quantified mathematically). On the other side, distinct from this objective reality, we have the "internal" subjective reality described as made up of experiences such as feelings, emotions, and thoughts. This is the difference, for example, between our experience of seeing the color violet and the range of the wavelength we identify as "violet". The modes of operation of these two "worlds" are dramatically different from one another, and an explanatory bridge that might effectively connect them has never been found. Descartes even claimed that such a bridge could not be found as a matter of principle. Modern emergentists insist that Descartes got this right (even if he got various details wrong): there is and can be no reduction of descriptions of our subjective experiences, of reports about our mental functioning, to the world of physical entities and reports on their behavior. This does not mean that modern emergentists think that our mental functioning is not ultimately the functioning of neural networks and no more, or that neural networks are not ultimately complex clusters of atoms and no more. This point bears emphasizing because it is often a source of confusion: unlike Descartes, modern emergentists are often no less ardent materialists than modern reductionists. They simply reject physicalism: that is, the confidence that every phenomenon that lends itself in principle to satisfactory scientific explanation will eventually be given a satisfactory physical explanation (Bunge 2003, pp. 146–147). So, for example, the modern emergentist points out that even the most complete electrochemical description imaginable of a neural network at the moment that it experiences the meaning of words read on a page or the taste of coffee sipped while reading them cannot replace a description of what it's like to have these experiences and therefore cannot explain the power of these experiences to provoke in us other experiences and thoughts. Understanding the physical pattern that governs the atoms that make up the person sipping coffee does not amount to an understanding of the experience of sipping the coffee and is therefore insufficient for an understanding of the power of this experience to provoke in us other feelings, thoughts, etc. The same is true of the very ordinary activity in which you are engaging at this very moment as you read and comprehend the meaning of these words: we have no idea how to reduce this experience to the realm of atoms and their properties. The reductive explanation, or at least that of traditional reductionism (which assumes that the world follows a simple mechanical —or at most statistical—regularity), appears to be obligated, in the final analysis, to exclude the entire realm of psychology from our world. The hope in this type of reduction is to replace all the psychological facts with biological facts, and then to replace the biological facts with chemical and physical facts (eventually dropping the chemistry as well).

We are left now with the task of introducing briefly two more clusters of puzzling emergence problems. These will play a central role in our attempts in subsequent chapters to understand the unique nature of communication as a distinct area of practice and study. From a very general point of view we can say that these clusters contain the most persistent problems of emergence that we are aware of. The first cluster, which I will now mention only very briefly, encompasses various aspects of the problem described in general terms as "the emergence of life from matter". Since the individual components of a living cell (e.g., the DNA molecule) are not in themselves entities to which we typically ascribe life, we do not know how the living cell emerged (and emerges) from them, in all of its various functions and patterns of growth, since in themselves they clearly lack these properties. Thus, we need not even posit the existence of a highly developed consciousness like our own to be convinced that we are in the presence of a most substantial problem of emergence. Organisms explore their environments, whereas elementary particles clearly do not. The issue involves a whole cluster of problems, all beyond the scope of this introduction: Biology, in all of its various branches, is a highly rich source of fascinating problems of emergence, the vast majority of which remain unresolved and most of which, unfortunately, we cannot even pause to introduce here. Ask yourselves, for example—just to tickle the mind—how the sexually reproducing organism emerged from the organism that reproduces by cell division; or how the organism that feeds on other organisms emerged from the organism that feeds on sunlight. To date, we have no satisfactory comprehensive reductive answers to these fascinating questions and many others like them (and to a large extent, only a complete solution of each of these problems will settle the question of whether or not it was fully solvable in the first place).

The second cluster of emergence problems that we will mention here is one that will concern us again in the final part of this book because it involves the subject matter of many scholars of communication in the narrow sense of the term (the sense that designates the core social sciences and their interconnections). This cluster includes the emergence of all the complex institutional dynamics that make up our culture as we know it, or in other words, the rather independent "life" that these institutions can be said to take on. Despite owing their continued existence to the power and decisions of individual human beings, these institutions display qualities that we do not know how to explain properly, since they are clearly not deducible even from the best imaginable knowledge about the motives of these human beings as individuals. Louis XIV could declare with his famous bravado that he and the State were one and the same, but anyone can see that France survived Louis' death, or that France but not Louis can develop, say, a galloping inflation. Economists point out that galloping inflation (for that matter) sometimes occurs even when the individuals involved in its creation all want to avoid it; how, using only the tools of psychology, can we explain the behavior of an institution that brings all its individual members to a point that is the very opposite of the one at which they had intended to arrive? Is such a reductive explanation even possible?

Sometimes we believe we know what information we are lacking and what kind of hypotheses about it we seek. For example, in the case of the inflation that no one desires, the economist might look for some simple factor that restricts the communication between the various economic parties involved in bringing about the inflation, or that restricts their access to certain kinds of content (for instance, information about the intentions of other "players"), etc. But oftentimes we do not even know what mechanism we are (or should be) looking for. This is why, for instance, no one knows how to predict the stock market's behavior in a year from now (or even tomorrow).

One particularly popular example of this kind of puzzling institutional emergence grows out of Émile Durkheim's classic analysis of suicide rates among Protestant and Catholic communities (that is, among communities with different levels of social unity and indeed with opposing ideologies of social cohesiveness). Durkheim found that even the most intimate of psychological decisions—the choice to take one's own life—does not take place against the backdrop of and due to mere personal despair but must be understood against the backdrop of and due to the level of social unity: it is grounded in the structure of the community of which the desperate individual is a member. The Protestant community, he argued, has lower levels of solidarity and integration, and its basic ethos is less concerned with fostering a sense of comradery between its members: its various institutions are less involved in the lives of its members and encourage them to favor self-realization over communal bonds. This increases the existential sense of detachment of the community's members, according to Durkheim. He argued that the suicide rate in such a community is significantly higher than its rate in the integrated and consolidated community, and that this high rate was a direct result of the weakness of the Protestant community's social bonds. If Durkheim is right, then no explanation that relies exclusively on the personal psychology of individuals in this community, no matter how complete, can account for its suicide statistics. These statistics, according to him, represent not a mere collection of private tragedies but rather a symptom of the structure of the community—a social structure of which the individual tragedy is but an accidental expression. This means that our psychological knowledge will always be lacking if we do not first complete the science of sociology. And since our social behavior is undoubtedly influenced by our aspirations as individuals, implying that we cannot explain the conduct of a given society without a deep psychological understanding of the motivations of its individual members, we appear to have arrived at a methodological impasse: our psychological understanding depends upon our sociological understanding, and vice versa. A purely psychological explanation of the phenomenon Durkheim studied is not unlike a pile of pathology reports explaining why each one of Job's poor children died, with no mention of the broader cause that had exposed them in the first place to the tragic circumstances of their deaths. Something is lacking: with respect to the phenomenon that Durkheim studied, the missing piece is a satisfactory sociological analysis of the loose social structure that caused members of the problematic community to develop feelings of a lack of purpose and existential horizon in the first place. But how do we go about studying the structure of a community? And how can we divine the centrality of this factor when all we are observing is the despair of the isolated individual, or of the individuals in isolation? What is lacking in the isolated reports on the deaths of Job's children is a piece of knowledge *on a different scale* altogether, namely God's decision to destroy them, the reasons for this decision, etc. In the absence of any preexisting knowledge regarding the vital relevance of a fact of this magnitude, it is hard to imagine how we might come to infer its existence.

Up to this point, everything I have said is independent of any particular view about the possibility of ever solving the thorny problems of emergence mentioned above. Reductionists and emergentists alike agree that these problems are very challenging indeed and that their solution is not imminent. For the purpose of our present discussion, we need to remember only that if we do not presently know how to remove all the conundrums of sociology and economics using only the tools of psychology, and how to remove the problems of psychology using only neurobiology, and how to remove all the puzzles of biology (ontogenetic, neurobiological, and other) using a unified theory of physics, then we effectively acknowledge our inability, even if it is temporary, to resolve certain troubling problems of emergence. Emergentists think that these problems are unsolvable by their very nature, and therefore that we find ourselves at an interesting and decisive crossroad: we can declare that certain prominent aspects of reality are not amenable to satisfactory scientific inquiry and explanation—certainly not by the natural sciences (Descartes, for example, had famously suggested this; he argued, perhaps somewhat ambivalently, that psychology, that is a science of our immortal soul, cannot be a real science, let alone a natural science, since our soul is not, in his view, part of the material world); or else, we can embrace all the sciences that study the various aspects of reality as legitimate, but stress that they institute a splintered or multi-layered epistemological system in terms of their relations to one another. According to this view, certain central aspects of reality, which can be studied and explained scientifically within distinct disciplines, can never be fully unified by means of a reductive translation into a single "basic" science, for the simple reason that each science studies different kinds of systems. This option was proposed over the years by many psychologists, among them even Freud, who argued that psychology was of course a full-fledged science but not one that is fully reducible by the theories of physics and chemistry. According to Freud, who was a materialist as a matter of course, even if the reductionist project were to be completed, we would still be left with not one but two basic sciences—physics, and psychology. A similar view, elaborated and enriched by further scientific layers, is proposed by Mario Bunge, one of the leading contemporary proponents of emergentism. He too is a materialist as a matter of course. But Bunge's position is much more complex and far-reaching: he holds that reality itself, and not just the sciences that study it, is made up of various systemic layers. In other words, reality, in his view, is layered not merely because when we study its many aspects we are forced to use approximations of markedly different orders and scales, but because reality itself is layered by virtue of the objective existence of systems. Bunge's position entails many difficulties that need not concern us here, and may well be ultimately inconsistent, but I will not explore it further. I recommend his beautiful and problematic writings to anyone who is intrigued by the problem before us.

One way or another, emergentists suggest that we should try our best to find ways of enriching our scientific explanations so that they will not be based solely on reductions to a more basic science. Such new modes of explanation (theories of emergence) would allow us, perhaps, to bring some measure of unification to our scientific worldview after all. But true unification, according to emergentists, will forever be an unattainable ideal. These possibilities, as we will see, are vital to our understanding of the status of communication studies and the prospect of studying communication as a science that unites various disciplines (which perhaps cannot be unified). Against these views, reductionists argue that there is and can be no scientific explanation other than reduction (and note that they, too, are tempted here to presume to know what knowledge will and will not be ours in the distant, unforeseeable future). To them, the expression "a theory of emergence" is simply a contradiction in terms, for emergence is at most a tag for our temporary ignorance, for everything that we have so far failed to reduce to a more basic science. Consequently, reductionists suggest that the only thing we can do in the matter of such dubious sciences as psychology, which assign us such dubious properties as subjective experiences and free will, is wait for the day when these sciences vanish of their own accord because they will have been replaced by genuine sciences, more basic and rigorous, such as, say, mechanics. When this occurs, we may well be described as walking automatons, more or less predictable, and the mistaken assumptions regarding the existence of our "inner world" and our "freedom" will be exposed as vague and deceptive metaphors, and removed.

Before we move on, I would like to point out a fact that is taken for granted nowadays by many people but which may be unknown to some of my readers: even if we set aside, as it were, all the wonders of life, consciousness and society, and even if we focus exclusively on those aspects of reality that belong traditionally to the worldview of the chemist and the physicist, who typically ignore life in all of its forms as a negligible phenomenon at the far-flung margins of the universe, still, physicists nowadays take it for granted that perfect prediction of the behavior of many different and complex physical systems is impossible. This is not merely a function of the limits of our knowledge (the fullest and most accurate knowledge imaginable) but also of the nature of the behavior of the components themselves. In other words, arguments for chaos in the behavior of certain complex systems have become an integral part of the theoretical world of physicists and of chemists as well, and not just of biologists, psychologists or sociologists. Thus, even the most ardent reductionists nowadays acknowledge that clear limits must be placed on the optimism of earlier generations regarding the prospect of one day truly explaining everything through a single, basic theoretical platform that will unify our scientific worldview as fully and as accurately as we would like. In earlier chapters, I mentioned the problem of predicting the state of the planets in the solar system, for example. But even in the case of a simple system consisting of only three bodies, like a system of two moons orbiting a planet or our planet and its moon orbiting the sun, we know that the slightest fluctuation in the behavior of one of them can result in significant deviations in the system's future behavior. Add to that the fact, now taken for granted, that there are certain known limits on the completeness and exactness of the knowledge that we will ever have about the basic constituents of many complex systems. To the best of our knowledge, perfect prediction of the future location of the planets is simply not within human reach. The more complex and unstable the system, the more limited our ability to predict its behavior. This is why, to give another example, despite the enormous resources invested in weather forecasts, meteorologists use embarrassingly vague terms like "a chance of local rain" and "no change," and why they are often wrong in predicting exact temperatures or humidity levels. No weatherman would wage his salary on his ability to give an accurate prediction (in non-vague terms) of the weather conditions in his vicinity two weeks down the line (let alone predictions for much longer ranges, like the weather on March 5, one thousand years from now). Therefore, even if we burry our heads deep inside the world of chemistry and physics, and even if we refuse to recognize the scientific legitimacy of the life sciences and the social sciences, a certain portion of puzzling emergence will remain an integral part of our excitingly teeming environment.

Bibliography

Bunge, M. (1996). *Finding philosophy in social science*. New Haven: Yale University Press.

Bunge, M. (2003). *Emergence and convergence: Qualitative novelty and the unity of knowledge*. Toronto: University of Toronto Press.

Kauffman, S. (1995). *At home in the universe*. Oxford: Oxford University Press.

Nagel, T. (2012). *Mind and cosmos: Why the materialist neo-Darwinian conception of nature is almost certainly false*. Oxford: Oxford University Press.

Rosen, R. (1991). *Life itself*. New York: Columbia University Press.

Rosenblueth, A., & Wiener, N. (1945). The role of models in science. *Philosophy of Science, 12*(4), 316–321.

Chapter 6
The Fundamental Problem of the Study of Communication

The prevalent attitude toward the puzzling problems of emergence described above is one of conscious disregard. I want to make clear that I am in no way criticizing this attitude. It rests on many good reasons as well as a fair amount of practical wisdom: many psychologists, for example, can thus go on exploring ways of alleviating the suffering of those who seek their help based on the assumption that the objects of their study are capable of exercising free will, while their colleagues from neighboring departments study the fascinating properties of neural networks and of course of individual neurons based on the assumption that the objects of their study have no such freedom, and each group can continue its fruitful investigation for many good years without heeding or even keeping abreast of the other groups' studies.

Yet it is no less pertinent to remember that these fields of study, which at present produce rather distinct levels of explanation of what sometimes appear to be almost distinct realities, are in fact not at all distinct. And the implication of this central fact is that the study conducted within each of these disciplines, however advanced, will never be complete before the disciplines are fully united in a single theory. The reason is simple: our mental life has an undeniable influence on the electrochemical behavior of the neural network that constitutes our brain as described by the neurobiologist and, of course, vice versa. Thus, so long as we cannot explain our mental states—for example, the moment of coming upon an original solution to a complicated mathematical theorem—using only the terms of a theory concerned with the electrochemical behavior of the neural networks from which these states emerged, we will not be able to give a full account of this reciprocity (this holds true even if we do not doubt—and most of us do not—that the reciprocity is indeed grounded in a physical identity and not just in a coincidental correspondence between distinct levels of existence). Note again that in the present context, "a full account" implies an explanation of the operations of our consciousness that would render all of our thoughts and psychic motivations wholly redundant. It means, in other words, that a scientist would be able to know what solution to the theorem the mathematician will arrive at without considering the mathematical problem at hand, indeed without any mathematical knowledge at all, based solely on his knowledge

N. Bar-Am, *In Search of a Simple Introduction to Communication*,
DOI 10.1007/978-3-319-25625-2_6

of the neural operations of the mathematician's brain and by consulting a suitable glossary of terms.[1] It means rendering redundant all of the mathematicians' thoughts and replacing them with a neurological, physical or other description of her brain activity; it means, for that matter, that the entire history of mathematics can be erased and retold through a reduction (with no loss of content!) to physics (plus a suitable glossary).

Well, the first order of business of this chapter is to point out that *unlike other researchers, who can postpone their treatment of puzzling emergence problems (at least for the short and maybe also medium term) in favor of ensconcing themselves in the temporarily secure confines of their traditional disciplines, researchers of communication cannot afford this luxury. This is because the study of communication in the broad sense of the term is the attempt to understand the mechanisms that enable the most puzzling cases of emergence*. (There: without deliberately trying, I have provided you with the closest thing to a definition of the field of communication that I am capable of producing. But please note that this is not at all a definition of the essence of the field but an attempt to characterize a cluster of problems with which we need to contend.) Thus, the pair of basic questions that seem to determine the possibility of conducting a general study of communication are: is a comprehensive theory of these puzzling mechanisms of emergence possible? And if not (since we will later argue for a negative answer), how can we nonetheless study them? I see this pair of questions as identical (with certain minor differences that I will clarify below) to these questions: Is a comprehensive theory of communication possible, and if not, how can we nonetheless study particular cases of communication?

The reason we cannot avoid confronting the thorniest problems of emergence when we wish to study communication is simple: "communication" as the term in its broad sense is widely understood is the name we have given to the mysterious modes of operation of the most prominent among the mechanisms of puzzling emergence mentioned above. It designates the manner by which a system emerges before us out of the mutual coordination between discrete units of matter. By virtue of their communication, the discrete units of matter emerge as the components of a system, and the system emerges as such.

This basic point is easily lost. It will be helpful, therefore, to revisit briefly the three basic examples of puzzling emergence noted above. Notice that in all three cases, the emerging system is enabled by communication, and communication is

[1]Even the logical possibility of such a glossary encounters difficulties that seem to me insurmountable. For if our prediction of the mathematician's proof is to be possible, we need to know in advance how to translate the neurological descriptions of her brain states into the language of mathematics. If the proof is indeed original, however, we will not know its full details in advance and accordingly will not know the original states it occasions in the proving brain. Consistent reductionists are therefore committed to the view that original thought is not possible. At most, they are to be regarded as new modular combinations of old bits of thought—but these do not amount to a discovery of new contexts. I will say more about contexts and their significance in subsequent chapters, where I will argue that the dispute between reductionists and emergentists revolves around the question of whether or not they can be eliminated from science.

just a name that we have chosen to designate a certain prominent aspect of the workings of the mechanism of emergence. Earlier, I mentioned the problem (or cluster of problems) related to the various factors that cause us suddenly to see matter as "living". These factors are admittedly rather vague, and some are in dispute among biologists, but that makes no difference for the present discussion. We need only take note of a fact that all participants in this debate take for granted, namely, that in identifying a certain piece of matter as an "organism" we are attributing to *it* an ability that underlies every act of communication: the ability to orient itself in an environment, *its own environment*, to get to know its own surroundings and respond to them. And yet, when we attribute this ability to an organism we are making use of a vocabulary, a set of distinctions, that we know not how to reduce to the language of physics, and that, moreover, we have no idea how to reduce without resorting to terms that the strict reductionist physicist, the one who aspires to a consistent reductionist worldview, must regard as blasphemy against his reductionism.[2] Biologists, for example, often and quite naturally speak about the organism's "internal environment", about its distinction from the organism's "external environment", about the organism's tendency to protect its internal environment against various fluctuations in its external environment, and even about "internal causes" for its form, functions, and the like (for example, the causes for the unique structure of its genome, the factors that give it one shape and not another, etc.) Conversely, we speak of the "external environment" of an organism, or of any other discrete organic system, and correspondingly of "external reasons" for its form and functions. This kind of external environment is a collection of possibilities and limitations on what we identify as the natural operation of the organic system, but the strict reductionist perspective, of course, does not

[2]Enthusiasm regarding various "solutions" to the problem of the emergence of life from matter continues to reverberate today among some reductionists, in light of various highly abstract definitions of life as an attribute of all systems that are capable of natural selection, heredity, slight variations in heredity, etc. If only we are granted permission to use terms like "function", "information", and "system", say those reductionists, we will be able to explain how it is possible for systems that adjust to their environment and evolve according to the principles of natural selection—that is, life systems—to emerge from the indifferent world of matter. This is very exciting of course, but what these eager scientists neglect to mention, and we shall try and explain in detail throughout this book, is that if we reduce the processes of the emergence of life from matter to various kinds of feedback mechanisms, we will not have removed their goal-directed or *telic* mysteriousness, but quite the opposite: we will have brought their emergent nature into sharper relief, and in terms whose non-reductionist nature is nowadays clearer to us than it was to our predecessors. A prominent example of an influential quasi-reductionist model of this kind is found in Gánti (2003) and is hailed the triumph of reductionism by Ginsburg and Jablonka (2015). They further argue that a similar model can resolve the problem of consciousness. In the coming chapters I will do my best to explain why this seems to me to be a misunderstanding of the problem at hand, and of the limits of reduction as an explanatory tool. "Function", "system", "information" and "feedback", I will argue, are simply not terms that the consistent and strict reductionist is allowed to use.

recognize such distinctions and indeed cannot recognize them as a matter of principle: it is designed to get rid of them in favor of a single level of description devoid of any "inside" and "outside" and expressing a single, final and exclusive level of operation (reality as it is in itself), which unites the individual organism and its environment and thereby abolishes their distinct existence. What the biologists calls a (more or less) stable living system, therefore, is invariably revealed from the strict reductionist perspective to be a heap of particles that cannot be distinguished from another heap of particles that encloses and embraces it (its environment), at least not without resorting to language that the strict reductionist is committed to describing as ultimately meaningless. The principles that hold together the organic system as a system, principles that allow us to identify it as a system and its constituents as constituents—i.e., its form and mode of operation—are thereby nullified and lost. Therefore, the strict reductionist cannot recognize the distinct existence of the organism as a system (with a system-specific environment). In fact, when biologists speak of organisms as systems that adapt to their environments they presuppose, whether implicitly or explicitly, that environments offer biologically meaningful action alternatives, and this seems to imply that organisms have purposes or goals (such as survival, heredity, etc.), and this kind of talk flies in the face of strict reductionism.

Words such as "purpose" and "goal-directedness" in particular raise a red flag among reductionists. After all, the scientific revolution of the past 400 years is so great largely because of the impressive intellectual achievement of replacing (explaining-away through reduction) traditional talk about mysterious goals and heavenly ends with theories that involve simple mechanical forces. In subsequent chapters I will devote more space to the topic of goal-directedness. I will argue that in a certain modest and unpretentious sense, ends survived even in the Newtonian worldview, and that they must survive in any picture of the world as a collection of systems and their mechanisms of operation. In particular, I will argue that we cannot study communication without positing at least a modest sense of quasi-goal-directedness. Indeed, I will argue that our very perception of a mechanism presupposes goals, albeit modestly in comparison to the traditional understanding, but nonetheless significantly so, since the strict reductionist cannot allow this presupposition. Without the functional differences that we detect between the system and its environment, let us note, it is truly hard to see how we could distinguish the system as such from its environment as such. In any case, until we elaborate our discussion of "goals" and equivalent terms, let us be as careful as possible in using them. Henceforth, such words will appear in quotation marks, and I will emphasize that they imply no more than the fact that the organism is perceived as a systematically organized bundle of fairly consistent responses to stimuli (responses shaped by natural selection), or in other words, a bundle of more or less predictable behavioral tendencies and patterns: each living cell, for instance, evolves according to certain patterns of growth, protects itself in various manners, and passes on "knowledge" (genetic and non-genetic) in a host of different ways. "Goal", therefore, for now, is just a way of saying that we detect this kind of marked consistency in the responses of all living organisms and that this consistency is attributed to them as systems, which is indeed responsible for our ascription of a certain direction to the development and behavior of these organisms and of their tendency to maintain themselves as unities, for example as homeostatic unities throughout some such developments. Indeed, it is often the case that we can predict fairly well the grown shape of an organism as a whole before this shape has "fully matured" (of course, we cannot predict this with certainty, nor completely, precisely because of the difficulties involved in reduction, but that is beside the main point here), and we can sometimes (not always, obviously) predict an organism's response to various challenges to its homeostatic stability. We sometimes also speak of the "function" of a certain organ in the body as a whole, and it is obviously only from the perspective of the biologist, the anatomist and their like, who regard the organism as a system and certain parts of it (organs) as distinguished by their function within the system, that the notions of "maturation" and of "function" carry any meaning. From the perspective of the reductionist physicist, this maturation and the functionality of an organ can play no role, certainly not one that would allow us to distinguish them from any other physical event, and even from the point of view of the reductionist chemist, the world is no more than a series of all-encompassing chemical moments ordered together as chemical causes and effects, actions and reactions, so that in each such chemical "moment" the organism is inseparable from its environment. We do not know what caused the emergence of the responsive consistency that we find in the organism and the directionality of its development, which allow us to identify the organism as a distinct system, but we have no doubt that they were shaped by natural selection. (Note that even talk of "natural selection" presupposes, as a matter of course, a distinction of the organism from its environment as well as a recognition of this environment as imposing selection pressure upon distinct organisms and upon distinct organic patterns or tendencies, and it is plain that from the perspective of pure physics all these distinction are quite meaningless, since their absence would in no way disrupt the physical description of the reality in which the organism and its environment operate.)

Some materialist reductionists argue that this responsive consistency of the organism as a system does not commit us to bring "goal-directedness" into the discussion, or that the talk of "purpose" even in the limited senses that we will consider later on is just a convenient (and ultimately disposable) way of saying things about the organism as part of the practice of biology. Others, like Norbert Wiener, the father of cybernetics, whose views will concern us later on in detail, argue that we cannot study communication and orientation in an environment even of artificial mechanisms, like the smart bomb or the thermostat, to say nothing of animals, without mentioning "goal directedness". Moreover, according to Wiener, and I will join him here in this view, a certain modest amount of "purposefulness" is a necessary condition for regarding the world as a collection of distinct physical objects that are markedly separate from their surroundings and can distinctly bring about well marked events in the world, i.e., can be the *full cause* of other events.[3] As we will see later on, even the central notion of a function in physics, despite a stubborn denial on the part of notable figures from among Newton's devotees, is itself modestly goal-directed. We need not go so far as to declare the actual (objective) existence of ends and goals; it is enough if we find that they are a precondition for describing the world as a collection of systems and their environment (without reaching any hasty judgments about the ultimate nature of the world, which is known to no one). And it goes without saying that no journalist or publicist or any professional working in the field of communication could perform their job without attributing to the particular entity that they are discussing its environment. This attribution is what we call contextualization, or placing something in context, and I will later discuss this act as rigorously as I am able, since I think that the ultimate question facing all communication scholars can be rephrased as the question "How do we share contexts?" In any case, a little further down the road I will also need to clarify the notion of a "goal" and spell out more specifically the boundaries of its legitimate and responsible use. For now, it is important only that we notice that even if biological "ends" do not have a fully rightful existence in the last and final theory of physics, everyone agrees that it is very hard to describe the living organism without them. The atom, after all, is not hungry, does not breath, has no survival instinct, no sexual impulse and no social needs. The laws that tell us how certain atoms join together to form molecules bring us no close to understanding, and no closer to explaining, how clusters of molecules can feel hunger or be overcome with an urge to survive or a desire for intimacy.

To simplify our preliminary discussion, let us focus for a moment on a single noncontroversial fact, and that is that *the organism is a system with a private environment* (henceforth, when I speak of the organism's environment I will be referring only to its external environment, although the same points apply with minor variations also to its internal environment). It is important to emphasize the *distinctly private* nature of this environment. It emerges with the organism by virtue of its being (or even its identification as) an orienting entity, a system that responds

[3]The idea of a full causal explanation as it is normally employed in science is undoubtedly the product of an *idealization*, and would consequently not withstand strict reductionist scrutiny, since in the final analysis, reductionism cannot recognize approximations and therefore cannot recognize distinct events that are the full cause of other events. As Rosenblueth and Wiener (1945, 1950) argued, without this vital approximation we are left with a notion of the entire universe as a single amorphous event that can perhaps be divided conceptually into past and future, but no more than that: the full causal explanation of an event (that is, the entire universe as a whole at a given "moment", whatever this term is supposed to mean here exactly) is the past state of the entire universe in the preceding moment (See also Rosenblueth et al. 1943). Of course, this kind of picture of the world rules out the possibility of any kind of scientific investigation and that is exactly the emergentist's point. Before Rosenblueth and Wiener, this view was powerfully articulated by Meyerson (1930): he exposed its foundations in the philosophy of Parmenides, which will concern us in the next chapter. There is, I contend, no fundamental difference between the idealization required for the concept of a complete cause and the one required for the concept of a goal or purpose: both are equally required for the work of the engineer and the communication scholar.

and adjusts itself to its surroundings. (This, indeed, is the original meaning of the term "subject", which designates something that is "subject to", that is "under the influence of" what lies outside of it.) The organism's environment needs to be distinguished from the objective reality that envelopes it and with which, to a certain known extent, it is impossible for the organism to come in contact: contact is meaningful (functional) mediation, and so its result is necessarily private. The individual organism's environment, then, is the sum of reality's possible influences upon its behavior, a mirror image of the sum of all of its private points of view, and these exist only by virtue of the workings of the organism's unique orientation mechanisms. In other words, the individual organism's environment is an outcome of its "goals", of the behavior patterns developed to attain them. The human body, for example, has in this sense the modest quasi-"goal" of maintaining a more or less constant temperature of 98.6 °F, and its environment includes possible challenges to achieving this.

To get a better sense of the distinctive subjective nature of the environment that we are discussing here, consider a standard ten-pound sledge hammer. The hammer is part of the human being's environment because it can, among other things, land on one's foot and cause a lot of pain and even real injury. But as Gibson and his followers have emphasized, it is not part of the environment of the *Escherichia coli* bacterium because even if it landed right on top of the *E. coli*, it would not stimulate the latter's orientation system: the hammer cannot provoke in the bacterium any real sort of response. Even the gravitational field, which of course surrounds all creatures large and small, is not part of the environment of the smallest known creatures: we know of no influence that gravity exerts on them. By contrast, many microscopic creatures are sensitive to movement that is not part of our environment as human beings—Brownian motion (the random motion of miniscule particles)—and in order to navigate their way, these microscopic creatures have to conduct themselves within this environment much like we would try to move around an amusement park filled with balls swirling about in constant and unpredictable motion. No one denies, then, that gravitation or Brownian movement are clear aspects of (objective) reality. But this does not mean that they form part of the environment of every individual organism that populates this reality. The organism exists and conducts itself in its environment and not in the objective reality, and the problem of the reduction of life to matter reveals itself here, inter alia, as the challenge of explaining the possibility of subjectivity (without resorting to such terms as "goal-directedness"). And so we see that the organism and its environment emerge because we identify communication among the organism's components, communication directed at preserving that organism and attaining its goals. These goals belong to the individual organism *as a system*, and strictly speaking, are not shared even by its components. This is exactly why an emergence problem emerges here. (And again: we do not know how to reconcile this fact with our confidence that matter *qua* matter has no goal or purposes at all; in and of itself it cannot even have a function.)

Allow me to summarize this point, since this is a mere preliminary note and I may have gone on for too long: the organism and its private environment, on the one hand, and the chemical-physical activity that enables it on the other, are two facets of the same reality. Like two sides of a coin. But since on one side of the coin we observe the organism operating in its (private, subjective) environment in light of known systematic "purposes", and on the other we find only an objective reality, with no internal/external distinction, no active/passive division, and with no goals, we do not know how to produce an explanatory bridge between these two sides.

In the coming chapters I will say much more about the challenges posed to reductionists by systems in general and by their goal-directedness in particular. Before I do that, let me recall briefly the other puzzling emergence problems that we encountered in the previous chapter and summarize the point I think they illustrate. The aim of the present chapter is not, of course, to resolve the dispute between reductionists and emergentists, but to clarify my claim that the fundamental problem of the study of communication is how to unify the main known puzzling problems of emergence under the umbrella of a single research discipline. Let us return then, briefly, to the problem of explaining the emergence of consciousness from the neural communication that takes place in our brain. We do not know what consciousness is, obviously. But no one nowadays doubts that it results from the complicated mutual coordination (that is, communication) among neurons. Consciousness and the inter-neural communication that enables it, then, are two facets of the same reality. *Here too, emergence is made possible by communication, and communication is just an everyday word we use to describe the fascinating operation of the mechanism of emergence*. Since the individual neuron is not conscious, we find ourselves unable to establish any kind of bridge between its properties as a piece of matter and its properties as a constitutive component of our consciousness. The solution to the enigma of consciousness, therefore, insofar as it admits of a solution, lies in understanding the unique nature of interneural communication as an emergence-enabling mechanism.

The same applies also to the social institution: it emerges from the mutual contact and communication among the human beings that know it and participate in it. For our present purposes, it is important to emphasize that *the social institution is not just the result of the mental and psychological make up of the individuals who establish it but also, in turn, shapes this psychic make up*: *it shapes the character and identity of the individuals whose joint action it organizes and from whose joint action it emerges*. The language in which we speak and think and the values that guide our lives are, of course, such social institutions: they not only enable our existence as a society and express our aspirations as individuals, they also participate in shaping our collective and individual character. This reciprocity is the fundamental observation that provoked the interest of the founders of communication studies in their various fields of expertise, from engineering and biology to sociology and behavioral economics. And I think that it is also the basic cause of all the challenges of emergence we encountered here, and the source of all the methodological difficulties involved in studying communication. It underlies our need, in the face of these challenges, for an explanation that cannot be entirely reductive. In the broadest terms, the problem with full reduction of the cases we discussed is grounded in the fact that, according to our best judgment, the communicative mechanism that produced (what we identify as) the system as a whole clearly affects the nature and identity of (what we identify as) its components as individual units. Therefore, *we can no longer understand the parts without understanding the system that shapes them, and* vice versa. To put the same idea in less abstract terms: the social institution and the people that constitute it are, again, two facets of the same reality, and since these two facets, as far as our understanding goes, exert a mutual influence, we find ourselves facing a weighty methodological problem, a problem of reduction that we do not know how to remove: sociological knowledge is a necessary condition for extensive knowledge of ourselves as individuals (as articulated in psychology), just as knowledge of ourselves (in psychology) is vital for a better sociological understanding… and the secret to understanding this complex reciprocity is interpersonal communication.

Here's a simple illustration. Consider a kindergarten, if you will. A rather common perception, which we will not dispute here, holds that every kindergarten has its "prettiest girl" and "strongest boy". Sociologists typically regard such archetypes as miniature social institutions. It is important to stress (especially for the sake of those who balk at the mention of such a crude example) that sociologists do not, of course, assume that these titles reflect empirical facts: the title "the prettiest girl in kindergarten" does not designate an empirical criterion for ranking the beauty of little girls, and sociologists who devote their time to studying such social institutions do not do so because they think that this or that social role is desirable or appropriate, or even defensible from an objective point of view. They merely note to themselves that this is a common social archetype, which also contributes to the shaping of identities: children seem to absorb such archetypes or stereotypes almost out of the blue, and this fact is particularly fascinating given that these roles have many gendered overtones and meanings, some of which we find disturbing. Children, after all, are not aware of the full extent of the chauvinist meaning of these archetypes when they embrace them and "allow" them to shape their behavior, their habits, and eventually also their personality. The fact that this narrative ends with the strong boy "winning" the pretty girl, like a prize, is a form of sexism ingrained in the system, that is, in the context in which our children develop.

And here's another interesting point to which the sociologist draws our attention: *the kinds of roles we are discussing are largely independent of the individuals who inhabit and realize them.* If the "prettiest girl" transferred to a different kindergarten or school, it would make no difference from the point of view of the group's dynamics: another girl would quickly take her place. The scholar of communication is impressed above all by the fact that more often than not, the newly anointed girl will gradually adjust her behavior and eventually also aspects of her personality to fit her new role, just as her unique personality will give the new role characteristics that are a reflection of her individuality. Thus, the role, which reflects the group as a system, directs the individuals and ends up influencing the formation of their identity (to us, perhaps, "the prettiest girl" is not such an important role in the process of identity formation, but it is a fact that beauty queens and supermodels direct their personality in light of their roles). And vice versa, of course. Roles direct individuals through widely accepted restrictions on their behavior, accompanied by social sanctions if these restrictions are violated. And this reciprocity, between the role that shapes the individual and the individual who chooses and shapes the role, is the fundamental phenomenon that fascinates communication scholars. The question they will seek to answer is how and under what conditions does this dynamic emerge from the individuals who constitute it, and what conditions enable it, in turn, to reshape their identity. From a methodological point of view, these scholars will therefore encounter a serious difficulty that those who ignore the problems of emergence do not sufficiently recognize, namely: how to go about studying a phenomenon that does not seem to submit to eliminative reduction.

To sum up: Communication is a phenomenon that transcends many fields and occurs right at the heart of the thorniest and most vexing problems of emergence. Moreover, it is a title (admittedly somewhat vague) that seeks to unite the various mechanisms that enable these cases of emergence. Therefore, if we wish to study the common features of these mechanisms, we cannot burry our heads in supposedly isolated disciplines, as many other scholars do, until the end of the days of science (as if such an end is even possible without confronting these problems): embarrassing cases of emergence are the stuff of which our discipline is made. The basic problem facing the student of communication, therefore, is how to go about studying such a complex and bewildering array of mechanisms of emergence.

Bibliography

Gánti, T. (2003). *The Principles of Life*. Oxford: Oxford University Press.

Gibson, J. J. (1986). *The ecological approach to visual perception*. Hillsdale NJ: Lawrence Erlbaum Associates.

Ginsburg, S., & Jablonka, E. (2015). The teleological transitions in evolution: A Gántian view. *Journal of Theoretical Biology, 381*, 55–60.

Meyerson, É. (1930). *Identity and reality*. London: Allen and Unwin.

Oyama, S., Griffiths, P. E., & Gray, R. D. (Eds.). (2001). *Cycles of contingency: Developmental systems and evolution*. Cambridge (MA): MIT Press.

Polanyi, M. (1968). Life's irreducible structure. *Science, 160*(3834), 1308–1312.

Reed, E. S. (1996). *Encountering the world: Towards an ecological psychology*. Oxford: Oxford University Press.

Rosenblueth, A., & Wiener, N. (1945). The role of models in science. *Philosophy of Science, 12* (4), 316–321.

Rosenblueth, A., & Wiener, N. (1950). Purposeful and non-purposeful behavior. *Philosophy of Science, 17*, 318–326.

Rosenblueth, A., Wiener, N., & Bigelow, J. (1943). Behavior, purpose and teleology. *Philosophy of Science, 10*, 18–24.

Chapter 7
Is There Communication in the Reduced World?

In the previous chapter, I argued that communication research in its most general form is concerned primarily with attempts to study common features of the most puzzling known cases of emergence.[1] As such, it draws together many of the major flashpoints of the controversy between emergentists and reductionists. Given the overriding popularity of reductionism among researchers, and especially in the exact sciences, this is undoubtedly a tense starting point. Yet we have no choice but to try to deepen our understanding of the reasons for this tension, since it embodies nearly all the methodological problems involved in the study of communication as well as most of the field's public relations difficulties. This is because communication scholars, as already noted, seek to unite aspects drawn from many different

---

[1]By this I do not mean to propose that we identify communication with the study of all known problems of emergence. This point was emphasized at the conclusion of the chapter "Emergence and Reduction", but it bears refinement here. Chaos researchers teach us that complex and unstable systems are common. Such systems are often puzzling for all kinds of valid reasons, but they are not puzzling in the same peculiar and distinctive way that the emergence of life or consciousness is puzzling. And they are not *as* puzzling. This may become clear when we compare the tasks of explaining the emergence of life and of consciousness with that of explaining, say, the emergent properties of a typhoon (the storm's direction, future speed, and even the moment of its formation). It may well be that there is no clear-cut criterion that allows us to regard the storm as any less an instance of communication than you or I. After all, even a typhoon can be described in certain known senses as having an "environment" in which it operates and which is a mirror image of our view of the storm as a distinct system. The willingness to define communication this broadly is not my own: indeed, it is my way of avoiding early undesirable conflicts with reductionists. For instance, in the opening paragraph of their classic essay, Shannon and Weaver (1949) use "communication" to encompass all the "procedures by means of which one mechanism… affects another mechanism", and it is obvious that such a sweeping definition was designed to deliberately blur the demarcation between what the physicist calls "interaction" and what we have here termed "communication", i.e., between what the physicist would describe as a system's "tendency" or "direction" and what we are here calling an individual's "purpose" or "goal". Weaver in particular expressed the hope that the theory of communication would one day be fully reduced to mechanics (and in the chapter devoted to Shannon's theory, I argue in detail that this hope was based on a cardinal misunderstanding of the nature of communication). Either way, it is commonly accepted that the central problems of emergence that we discussed here pose a challenge of a different order than that of the typhoon, or in other words, that communication is a particularly fascinating (if not always finely demarcated) subset of interactions. The reason for this strikes me as rather simple:

N. Bar-Am, *In Search of a Simple Introduction to Communication*,
DOI 10.1007/978-3-319-25625-2_7

fields, whereas the expert researchers in these fields have long ago become accustomed to carrying on like the inhabitants of parallel universes. The residents of these supposedly parallel universes often discredit one another's studies: many researchers in the exact sciences, for instance, regard the studies carried out in the social sciences as loose meditations with empty scientific pretensions, whereas researchers in the social sciences often complain that their colleagues in the exact sciences effectively void the human problems of all meaning when they regard the human condition as a collection of quantified pieces of data that are meaningless in and of themselves. Both complaints, as we are now in a position to observe, are merely symptoms of the deep bewilderment invoked in us by various problems of emergence as we go about trying to unify our splintered scientific worldview. And what troubles communication scholars (even if their interest in the field is not of a very broad theoretical nature) is that this distinction has become deeply entrenched even within their own department, acting as a fault line that separates colleagues: the "qualitativists" on the one hand, and the "quantitativists" on the other. And

---

(Footnote 1 continued)
the problems of emergence that we find most puzzling are those that stem from the overwhelming impression that a brand new property has emerged before us, such as the property of awareness to an environment. Since we have nothing like an absolute criterion for the identification of a property as "new", this should not be mistaken for an attempt to justify a position but at most as a way of conveying a vague impression (which may well be based on prejudice—for example, on anthropomorphic aspects of our consciousness).

Systems theorists tend to refer to the typhoon and the organism alike as "complex systems" but to distinguish the systems studied by communication scholars by describing them as "self-adaptive complex systems". The assumption, perhaps, is that the organism but not the typhoon adapts to its environment, and that this is the distinguishing feature that separates interaction from full-fledged orientation in a private environment. But this distinction is just as vague as the ones above and does not solve our problem. The air conditioner, for example, can easily be described as "adapting" to its environment when it adjusts its operation to the room's fluctuating temperature, as Wiener noted (and we could just as easily view the temperature as "adapting" to the air conditioner if we discuss it as a system). It is best, then, to drop the "self-adaptive" title, observing that the very word "self" is too context-dependent here, and to speak simply of "systems with an environment". This conclusion underscores the absence of any essential, objective, demarcating boundary that defines the phenomena studied by communication scholars: first, because the notion of an entity or organism adapting by itself is very vague, since even human beings (considered as early as Plato to be the epitome of self-propelled agency) are constantly influenced by various elements that make up their environment, and second, because even a markedly non-living system such as a storm has an environment in a certain loose sense, and this environment affects its "conduct". The storm is not, however, a subject in any standard sense of the word, so that the metaphorical nature of any talk about the storm's "point of view" or the storm's "needs" is very pronounced in comparison with talk about the organism's point of view or needs, and this is precisely the intuition that underlies the distinction commonly drawn between these two cases.

Let me sum up this point: the main claim of the chapters ahead is that the existence of (subjective) environments is inextricably linked to the attribution of goals (or problems, or challenges) to systems. This means that we are engaged in the study of communication wherever we are willing (!) to attribute goals (and thus an environment) to the phenomena before us (whether or not this willingness is backed up by a clear-cut criterion).

indeed, many quantitative researchers view qualitative research as a garbled attempt to circumvent mathematical ignorance, yielding unfounded blather that sometimes even accommodates data, with the wave of a pen, to fit a pre-formed ideology. The qualitativists, for their part, cast quantification as a futile attempt to invest numbers with properties that they do not (and cannot) possess, a grand pretense that numbers absent all interpretation carry independent weight, a pretense that the very choice of what to quantify is not already saturated with an agenda. Both parties exaggerate, since communication studies, as we have already seen, is not just one more discipline among many disciplines whose scholars hail from different schools of thought and pretend to inhabit parallel universes: the special nature of their discipline is the reason that it has branched off; the discipline's branching off is the very phenomenon that they seek to understand.

We have arrived at a key moment in the book: we are now in a position to summarize its main theoretical points. The present chapters offer a detailed attempt to elucidate the controversy between emergentists and their opponents. Following this exposition, I will argue that the basic bone of contention in this debate concerns the possibility of avoiding context-sensitive explanations in science. If they cannot be avoided—and I will try to show that this is indeed the case—it follows that we should examine closely what these contexts are as well as whether and how they fit into the standard scientific explanation (which ones should we appeal to, and why?). I will try to understand why it is that, contrary to the reductionist ideal, the many areas of study included under the broad title of "communication" (from biology to the social sciences) all assume, albeit as a temporary compromise but nonetheless as a precondition of their own existence, the necessity of such context-sensitive explanations. By contrast, I will also claim that we do not as yet know how to expand the reductionist conception of scientific explanation; in particular, we do not know how to incorporate context-sensitivity (which is by nature flexible and goal-directed) into a formal deductive framework. This is why later on in the book I will argue that *communication studies need to focus on developing a sensitivity to the multilayered (multi-contextual) nature of every phenomenon we come across (that is, to focus on the researcher's ability to construct alternative reductionist models of the same phenomenon alongside competing systemic models that explain the same phenomenon).* Part of what defines the scholar of communication is precisely his keen awareness of this multilayeredness and of the advantages afforded him by this awareness, over those who are not similarly aware and in particular over those who insist upon the reductionist ideal in its pure version. It is the scholar of communication who is distinctly cognizant of the methodological challenges that this multilayeredness presents before anyone who seeks a unified worldview (challenges that presently appear to be irresolvable as a matter of principle).

The most fundamental claim of the present book is this: w*e are engaged in the study of communication wherever we are willing to attribute environments, and thus "goals", to what we find before us*. Orientation and adaption, I will argue, amount to the ability to come into meaningful contact with one's surroundings. And this activity is necessarily goal-directed: it is carried out in light of and by virtue of given goals, so that any strictly reductionist description of this activity would fail to capture its spirit; if the reductionist denies the existence of goals, he thereby denies the existence of environments, and in the final analysis also the existence of any kind of orientation in the environment. He thus, denies the existence of systems. *Communication, on the other hand, is founded above all on the ability of various individual entities, or orienting systems, to share an environment with other such systems, and this presupposes the sharing of goals or orientation challenges (both with and against one another).* Communication is a mutual calibration by means of which various systems succeed not only in sharing the same environments but also in uniting as a meta-system to attain a common goal or to compete for its attainment. *The reductionist, then, whose goal is to ignore goals, ipso facto denies the existence of communication while at the same time necessarily partaking in it.*

To sum up: *reductionism is a fundamentally mistaken worldview that encourages a research plan that we cannot do without* (or, if you will, do not even know how to replace). *The aim of every research plan is to search for testable, context-free explanations. The reductionist worldview is mistaken insofar as it declares that a comprehensive explanation of this sort is possible.* In other words, the reductionist research plan expresses a scientific ideal of the first order, and its roots are indeed grounded in the only theory of explanation we possess (deductive explanation); but like every ideal, it is not fully realizable and in its fullest form harbors a veritable contradiction. *The reductionist ignores this fact, or else thinks that all context-sensitive explanations are ultimately reducible to context-free explanations (by including every possible context fully and explicitly within the theory itself). Herein lies his fundamental mistake: context-sensitive explanations are an integral part of the fate that entraps each and every scientist in his or her capacity as an individual engaged in the exploration of his or her environment.* This is also why communication studies will never be fully replaced by a future science that deals in the ultimate "building blocks", utterly free of context (whatever that expression may mean).

The chapters in this part of the book will necessarily involve fairly abstract discussions, which some readers might find challenging. They are included here for two main reasons. First, their conclusions (if not all the details of how they are arrived at) are vital to the tasks of this book: they are indispensable to our understanding of the complexity of any adequate explanation of the phenomenon of communication as well as to our understanding of its uniqueness and of the difference between this explanation and the reductionist explanatory ideal. Second, the scholarship that I am aware of is replete with interesting snippets of debates about the problems that concern us here; but I am unaware of any satisfactory systematic account of these problems, and so I see good reason to pursue it here.

Before we proceed, a few words of clarification about the reductionist world view. It includes an ontology (a theory about what exists, and the nature of being), a theory of explanation, a theory of meaning, and a research program for uniting these previous elements. I will discuss each of these aspects in the following chapters.

Reductionist ontology asserts that reality is necessarily composed of simple, eternal and basic items, or fundamental objects, and it describes their nature.[2] (Traditionally these basic and eternal items are called "substances", and it should be noted here that my use of the philosophically neutral term "item" is deliberate: it is designed to bypass various traditional controversies regarding the nature of the eternal and basic objects that reality is supposed to comprise; indeed even the question of their nature as objects is thereby successfully bypassed.) The reductionist theory of explanation asserts that all known phenomena are necessarily derived from these basic items. All of the objects we encounter in our lives, states the reductionist, are either basic items of this kind or else various configurations of these basic items (which are therefore not themselves basic items). Reductionists tend to say that these configurations, or "bundles" or "aggregations" of items, do not *really* exist—first, because unlike the basic items they do not exist eternally, and second, because the basic items, according to reductionists, are the cause of everything we see before our eyes. In the literature, these bundles or aggregates of basic items are typically called "epiphenomena" (that is, side- or even secondary-phenomena, which means effectively that these are non-phenomena… and thus are not deserving of an explanation). And indeed, one way of drawing out the basic controversy between reductionists and emergentists is through the question: can there be an explanatory contribution to the analysis of the world in terms of epiphenomena, in addition to its analysis in terms of the basic items? For it is important to note that emergentists argue that at least part of what reductionists call epiphenomena is more than an accidental and temporary aggregate of strictly basic items: in their view, there are real systems in the world. A system is a slice of reality whose explanation requires more than a description (however detailed) of basic items and their properties. In other words, the question in dispute here is whether every description of the world in terms of epiphenomena (and their apparently complex inter-relations) is already contained within the best and most detailed description possible of the world in terms of basic items and their behavior. We can also put the same point this way: is it possible always and in every circumstance to reduce explanations given in terms of systems to explanations that ignore the existence of a system-qua-system and refer only to the basic items that make it up? An explanation of the world in terms of basic items and their relations can be called here a "context-free explanation" precisely because it regards all talk about the system and its environment as ultimately superfluous. A successful reductionist

[2]The vast majority of scientists who are reductionists are also materialists—that is, they argue that the basic items that make up reality are investigated by physics (quarks, electrons, etc.). But it is important to note that there are also many reductionists who suggest that all known phenomena should be reduced to basic components of a mental rather than a physical nature. They are called "idealists". Leibniz, for example, who was one of the greatest scientists of all time, was such an idealist, as were the physicists Arthur Eddington and Erwin Schrödinger. I will not elaborate on their views since the metaphysical controversy between the materialists and the idealists does not concern us here. For present purposes, it is important only that we notice that materialist reductionists, who reject any talk of goals, find themselves in a very limited position vis-à-vis the basic phenomena that the researcher of communication finds so fascinating.

explanation, therefore, is an eliminating explanation insofar as it allows us to ignore the existence of systems as such and to treat them as no more than bundles or aggregates of more basic elements. An explanation of the world that combines several descriptive levels of reality as indispensable and irreducible components of the description of this reality, and in particular a description of reality in terms of systems and their environments, the relations between systems and their environments, the relations among various systems, etc. can be described here as a "context-sensitive" explanation. The dispute between the reductionist and the emergentist is about whether or not a complete theory of reality in which all context-sensitive explanations have been reduced with no loss of content to context-free explanation is possible.

Two key historical examples may help us here. The most famous reductionist ontology in the history of science is that of the ancient Greek philosopher Democritus. Democritus argued that the world is made up of tiny objects that he called atoms (*atomos*, to repeat, means indivisible). In between the atoms, claimed Democritus, is a void. The tree and the air conditioner, therefore, are not basic items and thus do not really exist: they exist only as Lego-like formations, or forms, made up of atoms and empty space. The tree appears to us like something that grows and dies, but the atoms that constitute it are eternal and unchanging, they neither grow nor die. The air conditioner appears to us to heat the room, but according to reductionists, there is no description of its operation that cannot in principle be replaced by a description of the operation of the atoms that constitute it. Therefore, in a certain known sense, there is no tree, no air conditioner, and not even a room being heated; there are only elementary particles surrounded by a vacuum. And indeed, this is the reductionist's explanatory theory: if and when we are able to describe and predict all the actions of the tree and the air conditioner while ignoring their existence qua tree and air conditioner and focusing exclusively on their nature as bundles of atoms, we will then have provided these actions with a complete reductionist explanation. Any other explanation of these actions—for example, a description of the tree's death due to lack of light or a disease, or a description of a malfunction in the air conditioner resulting from damage or wear of one of its parts —is tolerated by the reductionist as temporary and partial at best.

An earlier ontology that is no less important for our present concern (and to which Democritus' account is commonly assumed to be a response) is that of Parmenides. Parmenides argued that the entire universe is a single entity, a mammoth basic substance, solid and immutable, rather than many atoms separated by empty space. He held this view because he thought that the existence of empty space involves a logical contradiction (to justify this astonishing claim, he produced the first known proof in the history of thought!).[3] Thus, Parmenides would concur that the tree and the air conditioner do not really exist: according to his view, they

[3]For more on Parmenides' proof and its fascinating implications for the evolution of scientific thought in general and logic theory in particular, see Bar-Am (2008, p. 17–22), and Bar-Am and Agassi (2014)

are merely apparent slices of a whole that is not actually divisible because it is uninterrupted (i.e., solid or full), immutable, and lacking any "before" and "after", that is, lacking any time or change (Parmenides argued that the very act of carving up this reality requires us to posit empty space, and therefore is self-contradictory, and so impossible).

For our purposes, it is important to ask the following: how did Parmenides hope to bridge between the world-as-it-appears-to-us (in which there appear to be trees growing toward the light and birds chirping on their branches) and the solid and immutable substance that underlies it? The aim of science, after all, is to explain observed phenomena in light of what is described as the reality that constitutes and lies at the basis of these phenomena. We do not have an answer to this key question. On the contrary, it embodies the most general and most historically significant example of the problem of reduction in our culture: we are searching for the bridge between the multiplicity that we observe and the unity that we believe lies at its foundation, that is to say, for a reduction of the world of phenomena to reality. And yet, this kind of bridge, an explanation that removes all context from phenomena, is not possible in Parmenides' theory; indeed, his theory demonstrates particularly well why this impossibility is a matter of principle and not accident, since Parmenides declares openly that the tree and the birds do not really exist, and certainly not as discrete items, other than at most as imaginary and misleading slices of the immutable and uninterrupted whole that is the only thing that actually exists. How, then, are we to reduce the false appearance of the distinctiveness of phenomena? The most that we can do is explain why the false appearance is indeed false through a description of the truth as true. This kind of explanation is called "eliminative". (We thought, for instance, that there is an oasis before our eyes, but the scientist explains that it is an optical illusion and how it is caused.) But an eliminative explanation of the bird and the tree means just that: that the bird and the tree are eliminated, that they do not really exist, just like the imaginary oasis! Their existence, then, is not explained so much as it is cancelled, or explained away. If you and I, the birds, and even the trees and the rocks do not exist as discrete entities, and if change, motion, and time are mere illusions, as Parmenides appears to have held, then our private environment is also an illusion, and so is, obviously, our entire orientation in this environment. *All communication, therefore, is nothing but a misunderstanding of reality-as-it-really-is, according to the strict reductionist.* Reality-as-it-really-is contains no distinct entities (or systems), and therefore no communication between them.

And yet, this allegedly false world in which we apparently-exist as apparently-distinct-entities, and in which you and I are now apparently-communicating through writing and reading (and doing so in such a way that would be hard to write off as mere illusion), the trees apparently-grow toward the light and the birds apparently-sing—this rich world contains *all* of the phenomena that scientists are interested in and which they ultimately seek to explain! It turns out, therefore, that Parmenides' victory as a traditional reductionist is also his defeat: he eliminates (explains away) as an illusion all of the phenomena that he originally set out to explain. The scientist, it is commonly assumed, draws a distinction between a phenomenon and an illusion, and then attempts to explain the former as grounded in the basic reality and the latter as illusory. Could it be that the final scientific truth is nothing but the explaining away of all phenomena, nothing but the explanation that all phenomena are mere illusions? Parmenides has thus left us with the most fundamental mystery of the scientific endeavor: *the notion of the complete realization of the reductionist program exposes a contradiction in the thought that it is possible to unify the reductionist ontology and the reductionist theory of explanation*. Resolving this contradiction, as we will see, is a task that can always be postponed to a later date, and indeed is postponed by all reductionists with any common sense. This is the reductionist research program in action (it is how science actually progresses). But we must equally acknowledge that it is very hard to see how this contradiction will ever be resolved, and that there are rather strong reasons for believing that it cannot be resolved as a matter of principle.

And indeed, Democraitus' theory, responding to Parmenides, soon encounters very similar difficulties to those faced by Parmenides. This is, in fact, the most basic of the many different arguments advanced by various emergentists: that adding more basic items, more fundamental entities, and assigning them more basic properties, and in particular assigning new properties retrospectively in light of the discovery of new phenomena, new systems or new properties thereof, does not resolve the basic contradiction exposed by Parmenides but at most postpones it. For the reductionist ontology asserts that in the final analysis systems and their environments do not exist.[4] Instead of systems, it posits bundles of basic items and nothing more. At most, it adds to the basic items certain basic mechanisms whose purpose is to explain all forms of system-like properties (the legitimacy of this last move from the point of view of strict reductionism is a matter of contention, as we will soon see). In Democritus' account, this fact (i.e., the denial in effect of all systems and environments) is somewhat harder to discern than in Parmenides' since he tirelessly tried to address these difficulties in a host of unsuccessful ways, which we need not specify here. Suffice it to note briefly that Democritus—more or less unwittingly, simply by virtue of the fact that he allows for the existence of empty space between the atoms—allows his basic items to engage in certain interactions: they possess different and varied shapes and are permitted to move about in space, to quiver and vibrate, and even to bump into one another like tiny billiard balls. Thus, in his picture of the world, the problem of reduction evolves and takes on new forms, and perhaps is also momentarily dulled, but in principle remains unchanged: the question now becomes how to reduce the descriptions of the behavior of all the complex known phenomena to a set of rules concerning only very tiny, solid objects of various shapes that vibrate and bump into one another. Aristotle tells us that Democritus' atoms are like the letters of the alphabet: they are numbered, but

[4]The idealist reductionist does not concern us here, but we should nonetheless note that while he avoids the problem of the material existence of systems, he does face the problem of their mental existence, which is no less consequential or severe: if, for instance, the only existent *substances* are simple feelings, or elementary sensations, then we cannot explain the precedence of certain complex concepts (certain images of systems) over others.

can be used to write countless different books. This metaphor, in my view, is highly misleading and fails to capture the heart of the matter, since the vital question is what in Democritus' account distinguishes a bundle of atoms that we identify as a system and thus as functionally distinct from its surroundings, from an arbitrary bundle of atoms to which we do not attribute any environment at all. In other words, if the world is made up of densely packed letters, we need a criterion for distinguishing one book from another, a meaningful sentence (a combination of atoms that constitutes a system) from a meaningless sentence (an arbitrary combination of atoms). (And the criterion cannot be external to the world of atoms; it cannot be the result of a God-like, external perspective *on* this world.) What is it that distinguishes a random or "meaningless" bundle of atoms—say, an abstract ball of atoms that contains part of a tree branch and some of the air that surrounds it—from a bundle of atoms that appears to us to be special because of its functions as a system, like the tree, the air conditioner, the bird or human being? (Note, by the way, that even the modern molecule is not possible in Democritus' world, since of course the atoms that join together to form a molecule alter their internal structure as part of their consolidation, whereas Democritus' atoms are not allowed to undergo this kind of change.) The metaphor that likens atoms to letters, then, is useful only if we can supplement Democritus' description with a successful explanation of how it is that certain atoms, and they alone, persist alongside one another over time (like the letters of a book), as well as why it is that certain atom combinations are meaningful as a system (like humans or the air conditioner; that is, like the book, the chapter, the meaningful sentence, etc.) while other combinations are utterly meaningless (as, for instance, the atoms that make up the air conditioner if we pried them apart and scattered them all over the universe; like the fragments of a phrase or like readings that lack any meaning or unifying context).

Democritus, as I suggested above, tried to solve these difficulties in various and not necessarily compatible ways. For example, when he spoke about the distinctness of the living human being as an entity of sorts that is separate from its environment, he argued for the existence of a special unifying atom, namely the human soul, which in some sense constitutes the essence of man, in addition to the geometrical shape of the human body, whose relative constancy he also took to be an important part of our essence as humans. The first claim resembles the claim that the soul is aware of its environment by virtue of some sort of "tiny person" hiding within it who is in turn aware of his environment. This explanation is so plainly circular and faulty that it is hard to believe that the great Democritus, undeniably one of the most consummate and sober thinkers in the history of humanity, found in it any kind of intellectual solace. The second claim ignores the fact that relatively constant geometrical shapes also characterize many arbitrary combinations of atoms, like the abstract ball described above, the one that encompasses part of a tree branch and part of the air that surrounds it (or the shape that appears in Leonardo da Vinci's famous drawing as a symbol of the divine proportion concealed in the human body). Positing the existence of "the soul" as a wondrous and mysterious atom, therefore, can help us neither in delineating the human organism as distinct from its environment nor in explaining how the soul (or the "mind" that came to take its place in philosophical discourse) experiences its environment or communicates with its kind. On the contrary, the splitting of reality into material atoms on the one hand and mental atoms on the other only highlights and exacerbates the problem of explaining their interaction. What is it, then, in a world aligned more or less with Democritus' ontological picture that nonetheless enables the existence of complex entities with a private environment as systems distinct from their surroundings? How are we to explain the ability of certain entities to coordinate these private environments with other entities (thereby making the environments public to a certain extent)?[5] These are the problems that lie at the heart of the search for a comprehensive theory of communication, and they are precisely the problems that lay bare the shortcoming of Democritus' account. For anyone who eliminates the existence of systems-as-such will obviously have a hard time even articulating these problems as scientific challenges.

Once again, we find communication lurking stubbornly behind every central philosophical debate, however abstract and general in nature.

These difficulties have taken many interesting turns over the course of the history of science, which I cannot survey here, obviously. But the significance of this fact is

[5]These problems lie at the heart of the philosophy of Descartes, the most important and influential modern reductionist. Descartes' metaphysics is a particularly interesting testing ground for the problems I presented here because it provides a wonderfully lucid catalogue of the basic problems that his reductionist approach creates for anyone who wishes to acknowledge the existence of communication. Descartes sought to reduce the known world to two basic substances: a material substance, and a mental substance. But since substances, as we have seen, cannot communicate with one another as a matter of principle, he assigned responsibility for this communication to a third substance… an omnipotent God. Moreover, not out of consistency, Descartes divided the mental substance he had posited into many different mental substances, wishing thereby to ensure the compatibility of his philosophy with Christian dogma, which is founded on the eternal existence of distinct souls. This division did not sit well with his view of the mental substance as one by virtue of the common essence of all souls. The result was that the question of communication between the distinct souls posited in Descartes' theory, like the question of communication between mental and material substances, remained an unresolvable mystery (and once again the third substance, God, was brought into provide the miraculous solution). By contrast, the essence of matter, its fundamental property, Descartes identified as extension—in other words, the ability of matter to take up space in the geometrical sphere. In this, he yielded to the Parmenidean stance that the world of matter is an uninterrupted, undivided whole: for in the Cartesian picture, any particular item, whether a table or a person, is not essentially different from the medium that surrounds it (the essence of both, in the final analysis, is extension). Here, therefore, Descartes comes up against the Parmenidean problem of distinguishing the bodies from their surroundings—or carving them up, so to speak, as distinct slices of reality (the body and its surroundings have the same essence, extension, and therefore ultimately cannot be distinguished). It goes without saying that this problem is not satisfactorily resolved in Descartes' theory. His success as a physicist is due in large part to his avoidance of it. Incidentally, Spinoza's theory, according to which the natural world is all one indivisible substance of which *mind* and *matter* are different attributes, represents a brave attempt to tackle these difficulties, even if its implications remain somewhat vague. Here too, however, communication between individual items and their distinction from their environment is ultimately an unresolvable mystery, since for Spinoza as for Descartes, we are ultimately not at all distinct from one another (from God, who is also Nature). For Spinoza, the view that we are distinct entities is merely an expression of our relative ignorance.

that the reductionist theory has been repeatedly updated throughout history. And it continues to be updated from time to time. It is important to notice this dynamic and to emphasize that it constitutes the main reason for the vagueness of some of the debates between reductionists and emergentists, debates that presume to address the basic principles underlying the dispute and instead focus on matters of a merely local and contingent nature. Indeed, a regular revision and updating of the meaning of the reductionist terminology is the very crux of the reductionist program, a point that will be underscored and elaborated in the coming chapters. Thus, whenever human knowledge expands, reductionists stipulate the existence of new basic items and update the catalogue of their permissible properties. There is nothing wrong with this, of course: as Karl Popper emphasized, science is the fruit of our willingness to admit our mistakes promptly and correct them with a brave and open-minded spirit, and in this willingness lies its advantage (which many find so intimidating) over every other tradition (and particularly over those traditions that we consider sacred, since traditionally we treat the wisdom of our predecessors with respect, even reverence, and this inhibits the effective criticism of their views and the rate of their replacement). Such amendments of the reductionist ontology necessarily entail a reformulation of the reductionist argument, and with each reformulation, a slightly different boundary is defined between that which reductionists perceive as a mere bundle of basic items and that which emergentists regard as an emergent system with unique properties, with the debate turning more nuanced and delicate and technical the more sophisticated the reductionist ontology becomes. Modern physics has abandoned the traditional concept of substance, for example, but of course still posits certain basic items, elementary particles, as the fundamental constituents of the universe. The complexity of the debate increases, then, in step with the growing sophistication of the basic mechanisms that reductionists attribute to these basic items. As I already mentioned, it is hard to determine whether this move is legitimate from a strictly reductionist point of view, but even ignoring this issue, it should be rather easy to discern that the principle in these updated debates remains virtually unchanged from the time of Parmenides and Democritus. Newtonian mechanics, for example, is already not a traditional substance theory. First, it states that the basic items are ideal points that represent centers of gravity, but does not discuss the nature of the things represented by these ideal points, and this omission is very odd from the point of view of an ontological theory (what, in Newton's theory, are the sun and the earth in and of themselves?) Second, Newton's theory significantly expanded the scope of permissible interaction among basic items, inviting us in particular to attribute to the ideal points mysterious powers of attraction at a distance.[6] No wonder the legitimacy of this

[6]Plato and Aristotle claimed that the human soul is the only substance capable of moving itself, and indeed identified the operations of the soul, or human consciousness, with the power to move without being moved by an external cause, i.e., with the ability to possess a motive, or "mover" (which they identified with intention or purpose). William Gilbert, the sixteenth-century pioneer in the study of magnetism and electricity, argued that this definition applies also to the magnet and that we should therefore acknowledge either that the magnet has a soul or that other substances

move from a strictly reductionist perspective is questionable: it appears to be a retrospective re-introduction of a mysterious level of description that the reductionist was unable to reduce to the level of basic elements, under the cloak of basic properties. According to emergentists, consistency demands that the reductionist explain what patently appears like gravity at a distance as resulting from the existence of a more basic structure that clarifies why this mysterious interaction (which is a bit like telepathic communication between inanimate objects) is in fact not so mysterious. And indeed, Newton, the greatest scientist of all time, recognized this difficulty and did his best to remove it, but failed, as did many scientists who came after him, including his foremost follower, Laplace. In certain respects, this is the explanation finally offered by modern field theory, since what Newton's theory describes as a mutual attraction at a distance between objects whose nature is unclear (like the sun and the earth) is explained in field theory as an aspect of a structure. And so, it is indeed doubtful whether in this theory we can talk about celestial bodies as ultimately distinct from one another (just as it is doubtful whether we can ultimately talk about the tree as distinct from its environment, about the point at which the tree ends and the environment begins, and so forth). At most, field theory describes them as structures that are distinguished as separate entities in an unspecified ontological sense, for space is perceived there at once as containing separate entities and as identical with them. From a reductionist perspective, it is very hard to see how this duality might be consistently defended: you cannot split the baby and keep it whole; cannot attribute to the same reality two simultaneous levels of existence and ignore the question of how they relate to one another. The reductionist cannot ultimately accept this kind of dual language because the duality must eventually give way to a single level of description corresponding to a single, final and true level of existence. So, for example, the sun in Newton's theory must ultimately be either an ideal point in a Euclidean space or something else, and if it is something else, then its description as an ideal point in space is, in the final analysis, simply wrong; similarly, in field theory, earth is either a distinct entity in the field, or it is not; etc.

By way of summing up this point, it is worth mentioning one of Einstein's most famous followers, A.S. Eddington, who repeatedly cast the problem of reduction as the fundamental puzzle of science. He described brilliantly the difficulty embedded in the fact that the table in his room is made up on the one hand of dark, smooth wooden surfaces, and on the other hand of mostly empty space whose solid parts are colorless particles moving at dizzying speeds. This is, of course, the same

(Footnote 6 continued)

also possess the power to move themselves. The magnet, then, is the first material substance to which an unmistakable motive was attributed, that is, the ability to initiate change without coming in direct contact with anything other than itself. Newtonian gravitation is also such a property, though it is obviously not described as an "initiative". But the fact remains that all the fundamental forces in mechanics (and most prominently, attraction and repulsion) are based on mental verbs. This is not accidental: it reflects the kind of double language used to express a central aspect of the problem of reduction.

problem that has preoccupied us here, in a modern incarnation. Without getting into further detail, let me emphasize that to the best of my understanding, the bottom line of these debates is this: from the perspective that modern physics dictates to the modern reductionist, it remains true, as it was for earlier reductionists, that a system's distinctness from its surrounding reality is just the result of an *approximation*, that is to say, a momentary act of turning a blind eye in order to calculate a certain piece of datum out of the countless other aspects of this reality. An approximation, it should be emphasized (and this is the point that reductionists err in overlooking), is something that strict reductionist scrutiny will always and necessarily reveal to be an illusion or mistake (since an approximation is just another name for an inaccuracy that is good enough for the purpose of a particular local calculation but not for a more rigorous calculation, implying, in other words, that every approximation is an error from a more rigorous point of view. Reductionism, if it is consistent, aims to arrive at the ultimate and final level of description needed to explain reality as a whole, and therefore cannot ultimately make do with approximations). *The reductionist, then, ignores the fact that every idealization is an approximation, and that from a strict reductionist perspective, therefore, every idealization is an error.* (The population of the United States, for example, is approximately 317 million, but it is of course not exactly 317 million, and anyone who claims that it is exactly 317 million is wrong in his description of reality.) Eddington, by the way, brilliantly tried to avoid the impasse created by the fact that this kind of dual talk appears unavoidable and yet cannot be sanctioned by a strict reductionist with the argument that what is paramount in science is the calculation and not the ultimate nature of reality (instrumentalism). He even emphasized that the ultimate and final ontological interpretation of a given calculation is a matter of the scientist's private, personal opinion as a non-professional individual, an opinion that the scientist-qua-scientist should ignore (you may notice that with this, the reductionist effectively surrenders his materialism in favor of an instrumental worldview! And indeed, the most dogged reductionists are neither materialists nor instrumentalists but double-agents of sorts, swinging back and forth like able acrobats between these two incompatible positions!).Yet, the claim that strict reductionists find convenient to ignore and emergentists refuse to let them ignore is this: when the (ideal) day comes that we finally possess the most perfect and accurate theory possible, all of the approximations made only for the sake of local calculations are in principle supposed to be revealed as errors (big or small) in our description of reality. Whether these errors are more or less weighty, as errors, they must all be replaced by an accurate and definitive description that is not an approximation at all. With this (admittedly imaginary) achievement, it seems that all the systems that we perceive as distinct in our world, both animate and inanimate, will disappear, as will their environments. In this description, the communication taking place between us at this very moment is declared a total, uncompromising illusion, just as the substances of the world, motion, and time are denied in Parmenides' strange and solemn world.

In this chapter, I tried to explain in simple terms why reductionism harbors an absurdity that Parmenides had already articulated with remarkable clarity. Indeed, in the final analysis, the reductionist bears a surprising resemblance to his historical rival, the holistic philosopher, insofar as both are similarly hard put to explain the operation of individual entities in the world because both are ultimately committed to the view that only the universe as a whole is a full-fledged distinct entity (or what scholars of thermodynamics call a "fully isolated system"). This kind of ontological position (shared by advocates of holism and reductionism) is utterly useless from the point of view of conducting research, as emerges already from the legacy of Parmenides, since we cannot study the universe in its totality, as a single, indivisible object containing no distinct entities: without such entities, we have no phenomena, no regularities, and therefore no scientific puzzles and of course no scientific explanations.

For this reason, at least until we reach the final day and witness the complete and final explanations of our environment, I think that we are justified in ignoring the reductionists' unwavering conviction that they know in advance what general form the ultimate scientific explanation will take and that all problems of emergence are only apparent problems because they will be removed by this ultimate explanation, an explanation that apparently will also remove our very existence and the miraculous communication in which we are now engaged. As far as we can tell, the basic problems tackled by reductionists are here to stay, at least so long as we seek to explain the behavior of certain distinct objects as more than random bundles of purely basic entities. At least until that ideal final day, we cannot avoid acknowledging that the discussion of systems-as-such is a fundamental part of science—despite the fact that we do not know how to go about conducting this discussion properly.

Bibliography

Bar-Am, N. (2008). *Extensionalism: The revolution in logic*. Dordrecht: Springer.

Bar-Am, N., & Agassi, J. (2014). Meaning: From Parmenides to Wittgenstein: Philosophy as footnotes to Parmenides. *Conceptus, 41*(99), 1–21.

Descartes, R. (1985). *The philosophical writings of descartes in 3 vols*. Cambridge: Cambridge University Press.

Eddington, A. S. ([1928] 1948). *The nature of the physical world*. Cambridge: Cambridge University Press.

Meyerson, É. (1930). *Identity and reality*. London: Allen and Unwin.

Rosenblueth, A., & Wiener, N. (1945). The role of models in science. *Philosophy of Science, 12* (4), 316–321.

Schrödinger, E. (1969). *What is life? And mind and matter*. Cambridge: Cambridge University Press.

Shannon, C. E., & Weaver, W. (1949). *The mathematical theory of communication*. Chicago: University of Illinois Press.

Chapter 8
The Unbelievable Complexity of the Truly Simple

In the coming chapters we will try to deepen our understanding of the limitations of the reductionist worldview as the foundation of communication studies. This effort is worthwhile because outlining these limitations will allow us also to ask how, if at all, they might be circumvented. As an important introduction to all of this, we have to add to our discussion a brief but crucial note about the false promise embodied in the reductionist explanation. The most basic working guideline underlying all attempts to find reductionist explanations is to begin by *breaking down the phenomena* with which we are familiar *into their most fundamental parts*. This instruction is the main rule of method dictated to us by Descartes, who was the most important and influential of modern reductionists. His proposal was based on his conviction that once a system is properly broken down into its basic parts, the principle of its operation (its mechanism) becomes evident to us, (almost) of its own accord, since the most basic parts, according to Descartes, are also the simplest parts to recognize and understand, and grasping their natural mutual operation suffices in order to understand all the rest. Descartes' famous example is that of a mechanical clock disassembled into its *components* (the pendulum, cog-wheels, mainspring, etc.). His claim is that an in-depth observation of the components of the clock is enough to allow us to derive from them the principle of its operation, its clockwork, just as (so he said) an in-depth and careful observation of the axioms of Euclidean geometry allows us to arrive at a proof of a theorem that follows from them.

Within Descartes' proposal lurks a painful problem that many try to avoid or gloss over, and it is an incredibly simple problem even as its solution is nowhere in sight: after all, even if we do not know everything about every pendulum and wheel and all the other parts of the clock, we can easily appreciate the fact that there is a very large number of ways in which we can put them together, in light of different purposes; how can Descartes' theory of explanation divine and discuss all these many possibilities? To the emergentist it is obvious that Descartes' theory, or any reductionist theory that follows closely in its footsteps, is unable to discuss *all potential functions* of the clock's basic parts, or even of its components: *it is only after we have come to appreciate what a clock is, that is to say, only after we learn the purpose that man assigns to the clock, that we are able to assign its components their function.* When it comes to instruments, as I emphasized in the chapter on

N. Bar-Am, *In Search of a Simple Introduction to Communication*,
DOI 10.1007/978-3-319-25625-2_8

essentialism, it is the purpose that defines the instrument, (the purpose makes a certain item or bundle of items an instrument), and therefore different purposes define different conditions of success for different sets of combinations of parts. Moreover, even if we agree to build a clock from those clock parts that we have already recognized as clock components, and not some other components of some other mechanism, there would still be room to distinguish between the purpose of hanging a clock-like object on the wall for decoration and the purpose of telling time accurately; between different levels of accurateness in telling time, which dictate different criteria for a successful assembly of the components; and so forth. This point is highly important for our present concern: the need to know in advance the designated systemic function of the parts, and the need to rely on the mediation of something so vague from a physical point of view as the concept of a "purpose" is precisely what Descartes and other reductionists seek to avoid, since such a need implies that the reduction to the basic items is not truly complete insofar as the reductive explanation is not really based on the basic items alone (in what is known as a "bottom-up" explanation, i.e., an explanation that proceeds from the basic items and requires no further acquaintance with the complex phenomena). On the contrary: *if we did not know in advance what a clock is, i.e., what purpose is attributed to the particular slice of reality before us, we would not have the slightest idea how to disassemble it properly into its distinct "components" to begin with, nor any idea of what these components are, and moreover, what its "simple" parts are. For what the quasi-magical phrase "disassembling properly" means in this context is simply "disassembling so as to preserve in the disassembled components, to a distinct extent, their functional contribution to the particular system we wish to assemble*" (and indeed to do so, as it were, unintentionally, not by deliberate design). In other words, *the unspoken requirement is to disassemble the clock into its parts in a way that distinctly preserves their functions as components of a clock* and not of any other mechanism. Here, by the way, we start to see why the Newtonian "function" is nothing but a euphemism for "purpose": its goal is to legitimize the illegitimate goal-directedness in a way that is both subtly clandestine and circular. For it is by now clear that the function of a part is simply its contribution to the operation of the system as a whole in light of a given purpose (the very identification of something as a system presupposes this purposefulness). In order to appreciate this point it is enough to notice that even the smallest fragments of the clock are its parts, even after it has been smashed to smithereens with a hammer and welded together until they are unrecognizable as clockwork components. Yet it would be impossible to identify these as the parts of a clock necessarily, rather than of something else, or to reconstruct from them a particular clock without possessing advance knowledge of what that particular clock looked like (Rosen 1991, 20–23). *A component, then, is a part whose meaning (function) is already known, and this meaning or function can only be known in retrospect.* For of course, the atoms that make up the clock are also its parts: *an experienced clockmaker might be able to build a clock from its components without having first seen the complete assembled clock, but he could certainly not guess the complete clock from observing the bundle of atoms that made it up…. After all, from the atoms that make up a clock it is possible in principle (if only we knew how) to build not only countless different clocks but countless different systems with all kinds of different purposes, including instruments, about which we know nothing yet—neither their nature nor their purpose—because no one has yet invented them, to say nothing of the many instruments that we know but Descartes could not have known.* It follows that, against Descartes' own view but in full accord with his reductionist instructions, we do not know what constitutes the truly simplest part, such as the atom or any other part we choose to treat as basic, even if it is the familiar component of a clock, until we know from the outset all the possible structures in which it may potentially partake, including all the structures that humans may build from these simple parts in the future and of which no one has yet even dreamed! (Descartes understood the difficulty described here, and for this reason, his ontology does away with atomism, but to no avail from the point of view of his theory of explanation.) Descartes is caught here in a circular attempt to resolve the insurmountable difficulty of finding a firm and non-arbitrary foundation for his reductionism before the scientific journey has reached its (hopelessly ideal) endpoint. He is conflating unattainable ideals with reality. But it is just not possible to attribute to a simple part the sum of all its possible participations in all possible systems until we know these in advance.

Recognizing this important conclusion leads us to the main point I wish to make in this chapter, namely, that *reductionist ontology is in constant tension with the reductionist theory of explanation.* Specifically, the "the truly simple item" (of the reductionist ontology) is not identical with "the simplest item to understand" (from the point of view of the reductionist theory of explanation), nor, obviously, does the "truly simple item" designate something that is the easiest of all to identify and recognize from a "context-free perspective", whatever this phrase may mean (or even from an intuitive Cartesian perspective, to the extent that it is distinct from the supposedly context-free perspective). On the contrary: if reductionists are right, and everything that happens in the world does indeed happen by virtue of the truly simple, whatever its nature (and it is indeed very hard to see why they are wrong when this claim is considered in isolation), then from a methodological point of view, *from the point of view of the reductionist theory of explanation, so long as there is scientific progress, the "ontologically-truly-simple item" is probably the object that is farthest from our grasp and understanding in the entire universe, it is anything but "clear and distinct" since every possible scenario is already embodied in it, and we do not know most of these possibilities, and certainly not all of them, in advance and as a matter of principle.* We will most likely never know all of them. And this is the exact opposite of what Descartes imagined when he spoke of an intuitive grasp of "clear and distinct" elements. So long as we do not know what will happen tomorrow and what we will learn tomorrow, the basic building blocks of the world are the most mysterious and incomprehensible of all the phenomena that we may encounter.

And so, contrary to Descartes' view, it is precisely systems and their environments, and not the basic items, that present themselves before us in a relatively clear and direct manner, and that lend themselves un-problematically to research (this last fact is precisely what bothers the reductionist and piques the interest of the communication scholar). But the discussion of systems-qua-systems and of their environments-qua-environments necessarily presupposes *a context* without which we could never have conceptualized them to begin with: we could not even have identified the individual object as such—for instance, the clock as a clock—if this identification had not been preceded by a criterion that we do not yet know how to remove by using a perfect (i.e., context-free and final) reductionist description of truly basic items. Only such a criterion (a context) allows us to distinguish a clock as a clock from its environment as its environment. And this criterion, as is becoming gradually clear to us, is functional, that is to say, it is unavoidably goal-directed, since it unites for us a certain sum of aspects of reality on which we choose to focus our attention, a sum of aspects that is distinguished from the countless other possible aspects of that same reality. Only by virtue of this kind of selective isolation, which singles out a certain limited set of aspects of reality, does the clock emerge as a clock. Only by virtue of this selective isolation does an environment emerge from within reality.

Descartes' interesting analogy between the clock and Euclid's' geometry allows us to elaborate this point further. For the analogy suggests that *the difficulty we raised here must reappear also within deductive systems as such.* Today it is clear to us that mechanisms can be regarded as examples of recursive processes: they can be described as concrete materializations of abstract deductive principles. In particular, they can be described as a set of basic states (or axioms) and a set of operations that convert them into other states (or derivation rules). Now, Descartes unquestionably sensed this, but of course could not have realized it with clarity because truly abstract, or formal, deductive systems were not known in his time and his understanding of such systems was largely intuitive, if brilliant in his usual manner, very advanced, and even prophetic.[1] And indeed, his confidence—mistaken, as we know today, in retrospect—that the simplest items in our world would also be the simplest objects of inquiry, most likely rested on the fact that it was customary in his day to regard Euclid's axioms as self-evident, and thus as truths more clearly recognizable than any of the theorems derivable from them.

Thus, Descartes' analogy obscures the fact that no knowledge, however profound, of the axioms of a deductive system guarantees knowledge of all the theorems that can be derived from them (not to mention of the set of all the true propositions that are consistent with these axioms, as Kurt Gödel showed many

[1]The best known axiomatic system in history is Euclid's geometry of course. Non-Euclidean geometries were not known in Descartes' time. Euclid's geometry undoubtedly served as Descartes' model for the construction of scientific knowledge. But it was not the only deductive system known in his time: Aristotle's system of syllogisms (which preceded that of Euclid), and in particular the one implied by Aristotle's derivation of all the valid syllogisms from those of the first group was also rather well known, if not yet articulated in a way that underscored its nature as a formal system. Only modern scholars of Aristotle's theory (and most prominently, Lukasiewicz) could emphasize this aspect of Aristotle's theory unequivocally. Indeed, even the first fully formal articulation of Euclid's geometry belongs only to the time of David Hilbert. These facts clarify, I hope, just how far-sighted was Descartes intuition in this case.

years later with respect to the axioms of arithmetic). And since it is widely accepted by modern logicians (especially those of the Polish school) that the meaning of a proposition in a formal system is equivalent to the sum of all the propositions that can be derived from it, it is clear to us today that this implies that *we do not thoroughly understand the meaning of any axiom in any system worthy of our interest*.

All of this is, of course, much easier to see today, so that the above is not intended as a rebuke of Descartes and his philosophy. His reductionism was astounding for its time and its contribution to the scientific revolution was decisive. His boldness is truly unfathomable in contemporary terms: he hoped to base all the statements of logic upon (and explain them through or reduce them to) the law of identity (a = a), all of mathematics upon logic, all of physics upon mathematics and all of biology (excluding, perhaps, the biology that pertains to human mental activity) upon physics. In other words, Descartes sought to reduce (nearly) all possible human scientific knowledge to the law of identity which states that each thing is identical to itself. No less than that.

The circularity concealed in Descartes' proposal reveals a basic asymmetry between simplicity in the sense used by the reductionist ontology and simplicity in the sense used by the reductionist theory of scientific explanation. The reason for this asymmetry is our human nature, our ignorance as explorers of our environment. Because of this ignorance, it is our environment and not reality itself that we explore, since our environment is enabled precisely because we do not know everything about reality as it truly is, that is to say, because we live and act within contexts and not in a context-free objective space.

In the coming chapters we will see in greater detail how the reductionist research program handles this asymmetry. In a nutshell: whenever reductionists make discoveries, or even when they are forced to admit the mere logical possibility of novelty in their world, if they wish to incorporate this new possibility into their reductionist worldview they are allowed and sometimes obligated to update their ontologies and theories of explanation so as to enable them to derive the discovery from the updated definition of the simple items and their properties (or from the updated meaning they attach to the old terms). This, of course, restores order to the reductionist world: they get rid of the context by redefining the meaning of the term "basic item" so that it includes this context, i.e., includes all the knowledge that we possess to date, thereby abolishing the discovery as a discovery. In other words, reductionists thereby accord new and context-free systemic potential to the basic items, which are now once again described as "simple and fundamental" and as fully explaining the system's operation. (This is how the components of a clock become parts again.)

This kind of ad hoc correction is always possible, of course. And in itself, there is nothing wrong with it. But it is not possible in advance and as a matter of principle. As we will soon observe in greater detail, *reductions presuppose contexts* and therefore always "leave behind" at least one further unreduced context as a condition for their success. On this count, I suggest, the criticism of emergentists is accurate. The ideological disagreement between reductionists and emergentists thus may appear as a simple disagreement over the question of whether or not a final day can indeed arrive when we will no longer require these kinds of retroactive amendments: the reductionist is committed to the view that such a day will necessarily come, for otherwise he acknowledges that scientific explanation and the scientific picture of the world can never be complete as a matter of principle; emergentists point to a basic fallacy in the view that this day will arrive and therefore hold that it cannot arrive, as a matter of principle, and that the scientific journey is necessarily endless, a journey to improve our always-partial understanding of the reality that surrounds us. Either way, and even if we set aside this ideal disagreement, we must still contend with the question of how we ought to proceed until that final ideal day. We are, after all, trying to study communication today, not at the end of time. How can we do that, today? This is the urgent and fascinating question for which we do not yet have a satisfactory answer.

Bibliography

Bar-Am, N., & Agassi, J. (2014). Meaning: From Parmenides to Wittgenstein: Philosophy as footnotes to Parmenides. *Conceptus, 41*(99), 1–21.

Descartes, R. (1985). *The philosophical writings of descartes* (vol. 3). Cambridge: Cambridge University Press.

Rosen, R. (1991). *Life itself.* New York: Columbia University Press.

Chapter 9
Information in Context

In the previous chapter we described the ideal reductionist explanation, the one that is based solely on the basic items (and is supposed to explain everything we see before us "from the bottom up"), a "context-free explanation". What are contexts anyway? And why does the reductionist think that it is so important to overcome them and remove them?

The traditional reductionist answer to these questions is discussed in detail in the next chapter, but it is worth noting in passing here that *for the reductionist, contexts are perceived as a kind of vagueness or ambiguousness with respect to the precise reference of terms and thus with respect to the precise reference of theories.* Reductionism regards the removal of this vagueness as one of the foremost goals of science, since without perfect clarity there will necessarily be cases in which it is unclear whether or not our theories are true. Take, for example, the stock proposition "all swans are white". It is a known fact that this proposition was falsified with the discovery of black swans in Tasmania. Now, the reductionist way to fix our proposition in light of this refutation is to say that the original proposition is still correct, but only with respect to the swans of the northern hemisphere (which are, indeed, all white). Here the reductionist is fine-tuning the meaning of the term "swan" by splitting it into two terms. Yet the terms "swan", "hemisphere", and "white" are still imprecise, and therefore we will aim to clarify and fine-tune them in similar ways over and over again. We may discover, for example, a cream-colored swan in the northern hemisphere, requiring us to redefine the term "white" in a more precise manner, or else to re-define the boundaries of the term "northern hemisphere", and so forth. This we will repeat every time the need arises until, so the reductionist hopes, we arrive at an understanding of our proposition that renders it perfectly true with respect to all the basic items of the world—that is to say, until we rid ourselves of all ambiguities, all (inexplicit) contexts. The importance of this programmatic process for the reductionist, then, is that it brings us ever closer to a perfectly accurate theory, a theory that explains all known facts without exception and with no ambiguity. But, it is crucial to note that without such ambiguity, that is, *without conflating different elements in our surrounding (whatever these elements may be) by means of some reaction mechanism or other, without uniting them from a functional point of view (regarding, for instance, separate phenomena as "swan", one bundle of distinct elements as "threats", another as "sources of food",* etc.*), an environment of any kind is not at all possible!*

N. Bar-Am, *In Search of a Simple Introduction to Communication*,
DOI 10.1007/978-3-319-25625-2_9

Context, then, is a set of non-distinctions that have a functional significance relative to a particular individual or group of such individuals, that is, relative to an orienting system or a group of such systems. The lowest level of such non-distinctions can barely even be described as "embryonic abstraction" (because it is much lower than what we normally describe as an "abstraction" or a "generalization", since even creatures incapable of abstraction of course have environments by virtue of their response patterns to functionally similar stimuli). *Context, then, is the blurring of the distinction between different elements in one's surroundings, as they are united by response patterns. And the goal of the reductionist is to make all that is blurry clear and distinct: to explain away all environments as reality.*

The difficulties involved in this fascinating reductionist research program will be analyzed in detail in the following chapters. But it will help us first to try and understand again, in a more intuitive way and from a slightly different point of view, what contexts are. For the fact that *contexts are fundamentally ambiguities with a functional role* is not at all obvious, and can be hard to grasp at first. It is much more natural and common, after all, to think of context precisely as that which lends clarity to isolated situations, and this (mistakenly) appears to be opposed to the above reductionist description of context as the result of ambiguity or the blurring of distinctions between different elements. We also need to give a fuller explanation of the fundamental argument of emergentism: that the removal of all contexts as advocated by the reductionist is not possible as a matter of principle.

So let us pose the question once again: What are contexts and why is the reductionist so determined that they have to go? A good intuitive answer emerges from the brief chapter that we just concluded (which presented Descartes' circular justification of the possibility of a complete reductionist explanation). It became clear from our discussion there that every slice of reality—every basic item or group of basic items—can in principle *partake* in countless different objects and systems. The same atom, for example (assuming that the world is a sum of atoms), can partake in a stick, just as it can be part of a flower or of the mental activity that enables thought about a cinnamon cake. The stick, too, as a stick, can be part of countless objects and systems: it can serve as a lever, as part of the mechanism of a clock, part of the representation of a letter in the English alphabet, etc. (including, of course, systems that no one has yet dreamt of). Even the clock, in its turn, can participate in countless objects and systems, and so forth. Thus, we can say that context is simply the set of circumstances by virtue of which the countless possible functions of a given slice of reality are trimmed down to a more or less distinct function, or more precisely, to a limited bundle of such functions.

And here it turns out, that the first condition for this kind of narrowing-down of functions appears to be the existence of *a point of view* of some orienting system that observes the scene (or of a group of such systems that coordinate their focus on a particular aspect of their surroundings). The context is the result of this point of view, i.e., of the demand without which orientation in the environment is impossible: to unite under a single meaning (for example, a uniform response) various distinct aspects of reality (for example, to match similar responses to similar stimuli) while ignoring other aspects of that reality (all those aspects of objective reality that are not even "registered" by the orienting system). It is precisely this unification of different aspects of its surroundings, within the orienting system's point of view, that allows it to function in a consistent manner, to have action and reaction patterns and in particular, to function effectively in light of challenges. Thus, *contexts presuppose the orienting system and its environment.*

Imagine a stick, for instance. It can be perceived as part of a letter in the English alphabet if and when we use sticks—on the shore, for instance—to spell out a message. But it will be perceived as part of a message only by those who are familiar with the circumstances of using it as such, and as part of a message in English only by those who recognize the letters of the English alphabet. The chimpanzee, by contrast, will not identify the letters, but it will use the stick as an extension of its finger to catch ants: in this way, the chimpanzee enhances its ability to fulfill its needs, in its particular environment, and therefore the stick, for the chimpanzee, is not merely a stick but also an extension of its ability to summon food into its mouth. *Context, then, is a trimming down of the inconceivably rich diversity of any reality slice in light of a given set of functions* (a set that is not necessarily rigidly defined but which is nonetheless quite clearly distinct—certainly in terms of magnitude—from the inconceivable sum of all possible systemic functions of a given slice of reality, which is known to no one, as the list of all possible systems is known to no one). And so systems and their environments emerge from reality by virtue of this kind of system-centered point of view: because of it, the reality surrounding a particular slice of reality is perceived as its environment, and that particular slice of reality is perceived as an orienting system.[1]

---

[1]A note for the expert reader. This definition invites an infinite regression: the existence of context is founded here on the existence of an orienting system, which observes through this context and enables it. But this orienting system is in turn distinguished as a system only by virtue of yet another context, and so on to infinity (in effect, what we have here is an infinite regression of problems of orientation, which cannot be conceptualized without positing further problems of orientation that contain the previous problems or an aspect of them). We can put the same idea this way: there is no context without an observer; and the existence of an observer presupposes a meta-context. It follows that every context depends for its existence on a meta-context that envelops it. It seems to me that for emergentists, this fact poses no special difficulty since emergentism (unlike reductionism) allows for a denial of the possibility of a general and all-encompassing point of view, that is, of a final and complete über-context that unites all other contexts (by explaining them away).

In the popular literature, especially of the populist sort, there are those who seek to tether this argument against strict reductionism to the existence of a supreme observer of one kind or another, a divine spirit whose existence guarantees all contexts. Here I think that we can clearly see how deeply embroiled this literature is in a straightforward contradiction. *Context, as we clarified above, is the polar opposite of the idea that it is possible to contain all contexts in a single meta-context.* Therefore, such spiritualistic opponents of reductionism are no different from the reductionist, and in fact are a kind of impossible hybrid creature: "emergentist reductionists". The same title also describes those who think that it is possible to find a general theory of systems that will unify our understanding of all systems. This observation is, of course, not offered as an argument in the embarrassingly vague and futile debates about the existence or non-existence of God. Its purpose is merely to emphasize that from any possible human point of view, asserting the existence of an entity that functions as a ultimate-context of this kind involves accepting a logical

By virtue of this conduct, that is, by virtue of exhibiting functional tendencies, by virtue of its various "goals", the infinite number of possible perspectives on a slice of reality is narrowed down. This is what allows the slice to be perceived as a particular system, or as a particular part of a system, and not as the unlimited, infinite set of possibilities embodied in this slice—as in all the other slices—of reality.

In some respects, even seeing the stick as a stick is context-dependent and therefore the stick-qua-stick cannot be regarded as a truly basic substance, of course: for a tiny creature, the stick may constitute an entire habitat, and for a particularly large animal it may be so small as to be indiscernible and therefore not at all part of the animal's environment. McLuhan's basic observation regarding all media also relies on this context-dependence, and it accentuates it, since in order to know what constitutes an orienting system's environment—for example, the environment of a chimpanzee—we need to ask, according to McLuhan, whether and how it is able to enhance itself, that is, to fulfill its goals, by means of its surroundings. Media, it now emerges clearly, is always a relative term: it depends on the orienting system in question. If the stick carried no potential whatsoever for extending the function of a particular system, it would be meaningless to it, and would not be counted as part of that system's environment, as it would not have been perceived by it.[2]

Every slice of reality, then—and also, by the way, such abstract items as propositions, melodies, etc.—possesses in principle an infinite number of possible meanings because it can in principle partake in an infinite number of different systems, and in other words, belong to countless different environments of different orienting systems in different contexts. Thus, the same proposition can say entirely different things to different individuals, but its meaning changes also when it is situated within different stories, i.e., when its contribution as part of an overall narrative changes, whether the change occurs in the character of the individual to whom the narrative is addressed, in the individual's level of understanding, or in the story itself.

With this preliminary discussion of the meaning of the term "context" we have already touched upon one of the central meanings of the term "information" as we will use it here, if only as representing one end of a spectrum of terms that denote the phenomenon we wish to study, that is, as a limit case to which we have no real

(Footnote 1 continued)

contradiction. (Since this conclusion is gladly embraced by quite a few irresponsible individuals on both sides of the abovementioned debate, the dispute rages on.) Incidentally, the contradiction I described here is clearly and beautifully revealed already in Leibniz's metaphysical system, for example in the claim that all truths are perceived as tautologies and proven to be such from the point of view of God.

[2]We should note that the question of what constitutes a given individual's means of communication is one that must be posed from several points of view at once, for example, both from the conjectured point of view of the chimpanzee and from a more general point of view, on the surroundings that *we* perceive as human beings observing chimpanzees, that is, on the possible expansions of the chimpanzee's environment as they are conjectured by us as part of our understanding of its surroundings. Thus, "environment", "surroundings" and "reality" are distinct theoretical terms with distinct and separate meanings.

access: information is first of all the possibility of meaning (of whatever sort, in whatever form); it is *the potential for content*. As such, information is above all a basic aspect of reality-as-it-is-in-itself, and in particular of reality as we imagine it to be from an "objective point of view". ("An objective point of view" is an oxymoron, of course, but it is one that we cannot escape in describing this particular predicament: a point of view cannot truly be objective because, being a point of view, it is subjective. Similarly, an individual cannot reduce all contexts and remain an individual. This is the whole reductionist self-contradiction wrapped up in a single proposition—reductionists seek to unify solipsism and realism through an explanation that eliminates all their experiences as if they were nothing but pieces of information: it is thus the reduction of content to the potential for content.) However, as a matter of fact, reality as the scientist imagines it can become, and does in fact become, many different environments. All of this occurs thanks to many different contexts, i.e., through the indispensable mediation of systems and their varied points of view.

Henceforth, we will call information in the sense I just described by the name "absolute-information". We saw that it is not something that any particular individual ever faces or ever could face: philosophers call it by a name given it more than two hundred years ago by Immanuel Kant: "the thing-in-itself". Systems orient and conduct themselves in environments, they conduct themselves within them necessarily, and thus do not and cannot conduct themselves vis-à-vis the thing-in-itself, that is, within the realm of the pure potential of reality to become their environment.

Yet in the coming chapters, I will almost always use a more important sense of the term "information": a sense that I will call here "relative information" to contrast with the term defined above, and sometimes simply "information". I will define this term in detail later on, but this much I will say now: relative information is slimmed down content, it is the abstract representation of content after the context that originally gave it meaning has been lessened or trimmed.

It is worth nothing, by the way, that in the literature, the uses of the term "information" are highly varied and incompatible with one another. A list of competing definitions compiled by Schement (1993, pp. 20–30) provides some good examples of the contradictory uses of the term that prevail in the scholarship. Of course, my intention here is not to "correct" common uses, since every person is free to use any term as they please, as emphasized already by Plato (provided that their use of the term is understood by their interlocutors for the purposes of a given end and in the context of an adequately defined problem). We cannot require uniformity across an entire literature. My aim here, rather, is simply to help readers find their bearings in the labyrinth of different definitions, some of which clearly express confusions and misunderstandings regarding the theoretical relations that hold between content and information.[3] In particular, my

[3]Here is one example of such a confusion. Schement's list of definitions includes the highly influential definition offered by Hayes (no. 7), one of the fathers of information studies:

aim is to highlight the common and harmful confusion between relative information and content or meaning. This confusion lies at the heart of reductionism, as we will soon see, since the reductionist program is founded on the hope of reducing all contents to information, by means of formal systems, on the way to their ultimate reduction to absolute information (that is, to the aspired explanation of all environments as the result of reality-as-it-is-in-itself by means of the final and perfect scientific theory at the end of days).

Up until now, we discussed the orienting system and its environment, emphasizing that they are complementary concepts: the one cannot exist without the other. Perhaps we may say that the orienting system embodies a set of orientation challenges, and its environment embodies a possible set of solutions to these challenges (not necessarily successful). This is also why the existence of orienting systems presupposes the existence of goals: in the absence of the goal, no behavior pattern is meaningful. Purposes, or goals, can be built-in or acquired, or as we will later see, externally-imposed, or projected upon. The material of which the air conditioner is made obviously has no purpose other than the one attributed to it by the engineer and the user in their perception of this particular arrangement of matter as an air conditioner, but the air conditioner qua air conditioner does have an environment—unlike material qua material, which has no environment because it has no goal. And indeed, according to the reductionist, all purposes and goals in the world are ultimately projections. From a realist point of view it is tempting to say that it is by virtue of purposes and goals that absolute information (reality) is sorted into contents in the world of orienting systems and their environments, in other words, that reality is transformed into subjects and their experiences through the mediation and by virtue of contexts. But it is important to note that since orienting systems never conduct themselves within the realm of reality-as-it-is-in-itself, they are always operating in a realm that is somewhat purposeful or goal-directed.

So, we now see that context is a systemic-goal-directed aspect of reality, it is *telic* if I may say so, although perhaps not *teleological* since I do not mean to say that it entails any hierarchy of goals, and certainly not anything like a penultimate unifying meaning. Context is an aspect of reality that is imposed upon it by virtue of the existence of orienting systems, by virtue of their goal-directed conduct. It is by virtue of orienting systems that reality ("the thing-in-itself") becomes a particular environment for a particular orienting system or for a group such systems as they coordinated their conduct, their goal-directed behavior (and it matters little, in this respect, whether such a goal emerged with them, or was built into them or imposed

(Footnote 3 continued)

"'Information' is data produced as a result of a process upon data". The word "data" appears here, perhaps deliberately, in two very different senses, and without forewarning: the first is data prior to its processing, the second—after its processing. This is, at bottom, a confusion between a term in an object language and a term in a meta-language. This confusion is a key move in the more basic reductionist confusion between content and relative information, as I will describe below. It forms the basis for the illusion that relative information is ultimately reducible to absolute information, i.e., that content can be fully explained by means of an objective description of reality.

or projected upon them by other such systems). In other words, *the emergence of contexts is just as problematical for the strict reductionist as the emergence of goals, for strictly speaking, goals and context are inextricably entwined.*

So much for our discussion of context from an intuitive point of view. Let us now return to the reductionist who regards contexts as covert ambiguities that can always be removed—as opposed to the intuitive conception of contexts as enabling orienting systems and their environments, and as impossible in principle to remove (explain-away by means of reduction) fully. It is apparent that for reductionists, context is a handicap or limitation that cannot be tolerated: their task is to see all systems as collections of basic ontological items and nothing more—in other words, to explain-away the goal-directed behaviors of these systems, to explain them by reference only to the basic items. Now, this task can be carried out only if we are able to reduce all contents to absolute information, that is, to encompass all contexts in a single context-free description (of the basic items). The hope is to render redundant all talk of orienting systems and environments by providing a comprehensive description of the basic items and their (essential) properties.

But how do contexts appear from the point of view of the reductionist, that is, from the point of view of someone who wishes to deny them? For if a "point of view" is but another name for what an orienting system experiences in a given context, and if the reductionist denies the actual existence of systems and their environments, then it seems that the role of the reductionist is (in the final analysis) to explain away the existence of all contexts. Though this may seem surprising, there is in fact such a thing that deserves to be called "the reductionist point of view" even if the reductionist program is based upon the constant, principled denial of its existence. This is a rather subtle point, and in any case, my own impression is that a significant part of the popular and professional literature alike gets it wrong, almost systematically. In order to cast off these confusions and to understand why they are systematic, it will help us, rather than delving deeper into this literature and surveying its mistakes, simply to try to lay the ground for a straightforward understanding of the problem set before us. We need to understand what remains of content after its translation into a theoretical framework that the reductionist accepts as legitimate. And we need to understand whether what remains allows us to study communication properly. These are the questions that will preoccupy us in the coming chapters.

It seems to me that the best way to go about trying to answer these questions is to consider the reductionist theory of meaning. In the scholarship, this theory is known by a slightly intimidating name: "extensionalism". Since the discussions surrounding this theory of meaning tend to be highly technical and long-winded, they alienate many potentially curious readers, which is of course no one's deliberate intention. In the next chapter, I will try to avoid these theoretical complexities as best I can, and those that remain, to present in a user-friendly way. Getting to know extensionalism is worth the effort, because it helps us understand how contents and contexts are perceived from the point of view of those who want to deny their existence. This, in turn, will allow us to grasp the main argument of the subsequent chapters: that despite the many local victories of reductionists in their righteous war against contexts, it is clear that they will never be able to remove them once and for all. These, therefore, are also the limits of their righteousness.... We have a fascinating case here: a locally justified battle, which from a general and principled point of view is fundamentally mistaken and hopeless.

This conclusion is significant: it means that the insistence of emergentists upon the intrinsic existence of problems of emergence in our scientific picture of the world, including the scientific picture of the future world that we do not yet know, is justified, and will remain so in the future, so long as our methods of explanation and proof do not change radically (in a way that we cannot even imagine at present). Therefore, the significance of our present theoretical discussion is of interest not only to philosophers and their like: it bears directly and immediately on the limitations (which hold, to the best of our knowledge, as a matter of principle) of studying communication, and more generally on the limitations of our entire scientific picture of the world.

Before turning to the discussion that explains these claims in detail, let me repeat, I am no enemy of science. But I am convinced that, despite all of the reductionist's protests, we cannot avoid the conclusion that goal-directedness is indeed part of the very structure of our picture of the world. Reductionists deny this. *Acknowledging the unavoidable existence of problems of emergence as a basic fact of our reductionist worldview seems to me to be a first step toward understanding the complexity of the challenge facing students and scholars of communication* and a crucial stage in the attempt to understand the possible ways of addressing these problems. Because this point is so sensitive, I am devoting to it more space and closer attention than may seem warranted in an introduction that aspires to be reader-friendly to beginners in the field.

And if I am already pleading your indulgence, I hope you will allow me one more brief clarification before we roll up our sleeves and get down to the fine technical details. In my descriptions of the reductionist position, I return in different versions to two basic claims. The first is that the reductionist methodology is absolutely vital to science (as we will see in the next chapter, it is the very foundation of scientific explanation). At the same time, I also emphasize that the full realization of the reductionist methodology is not possible, and were it possible, would amount to the negation of our ability to understand our environment, to say nothing of our ability to know our way around it. For knowing one's way around one's environment, as we have seen, consists in the constant narrowing-down of reality's infinite capacity for content making to a limited set of meaningful experiences circumscribed by a certain functional point of view. Without this narrowing-down, no orientation is possible, and without orientation, the narrowing-down is not possible.

My final point of clarification, then, is this: it is hard to see, at first, how progress toward a goal can be such a welcome and vital achievement when its full attainment would amount to a stinging defeat (as well as the loss of everything achieved during the gradual progress toward that goal). On closer observation, however, I hope you will agree with me that this circumstance should no longer surprise us. We all know many such examples. In fact, this circumstance is already suggested by the common meaning of the term "ideal" which denotes a noble aspiration whose full realization is not actually possible. The reductionist ideal is no exception to this rule—and that is the whole of emergentism in a nutshell. One example, which happens to be pertinent to our discussion, comes from looking at the classical theory of rationality. This theory is based on the idea that rational choice is a choice that is optimally justified (relative to given circumstances). And yet, all classical theories of rationality take for granted the assumption that the optimal, or ideal justification, is the proof. But a proof forces its valid conclusions upon us, and does not offer us their validity as an option or a matter of personal choice. It follows that optimal rational choice is no choice at all: it is obedience, even blind obedience, to data.[4]

Back to reductionism, then. We have seen that a successful reductionist explanation simplifies and abstracts, or generalizes the phenomena that we encounter. But the ultimate reductionist explanation (what the pre-Socratics had already called "the unity in diversity") is the reduction of all phenomena to the Parmenidean "one" (or to the Cartesian "a = a"). So it is not surprising that such an explanation, were it possible, would indeed simplify our understanding of phenomena to the limits of our ability, but at the same time would annihilate everything that makes them distinct phenomena to begin with—thereby obviously vitiating its function as an explanation.

Bibliography

Bar-Am, N., & Agassi, J. (2014). Meaning: From Parmenides to Wittgenstein: Philosophy as footnotes to Parmenides. *Conceptus, 41*(99), 1–21.

Polànyi, M. (1970). Transcendence and self-transcendence. *Soundings, 53*(1), 88–94.

Reed, E. S. (1988). *James J. Gibson and the psychology of perception*. New Haven: Yale University Press.

Reed, E. S. (1996). *Encountering the world: Towards an ecological psychology*. Oxford: Oxford University Press.

Schement, J. R. (1993). Communication and information. In J. R. Schement & B. D. Ruben (Eds.), *Between communication and information* (Vol. 4, pp. 3–33). New Brunswick: Transaction Publishers.

[4]Here is another example from a different field: political theory. The basic ideal of every liberal democratic government is the maximization of personal liberty and minimization of state intervention. It is obvious that the more progress we make toward maximizing this personal freedom—without, of course, infringing on the rights of other members—the greater will be the liberal achievement of our government. Now, the ideal height of personal freedom is our complete deliverance from any reliance on the governmental institutions that restrict us. But this is anarchy. Anarchy marks the total defeat of the liberal government (and by extension, also does away with a large part of our ability to realize our basic rights—for example, the right to be protected from theft and robbery). It follows that the liberal ideal is anarchy, but anarchy is the end of liberal government: therefore, freedom circumscribed by the law is much more freedom than absolute freedom, which is hardly any freedom at all.

Chapter 10
The Reductionist Point of View—Extensionalism

The goal of the coming chapters can be summed up as follows: (A) To explain why, in the final analysis, the reductionist claims that (meaningful) content is nothing more than absolute information. In other words, the reductionist claim is that since meaningful interaction is nothing more than an aspect of reality (as it is in itself), it must be describable in principle as lacking any context, even if we do not yet know how to articulate it as such. (B) To explain why it is that, even though it is always possible in retrospect to reduce a certain limited piece of content to (relative) information, the hope to reduce all content to information (the kind of wholesale reduction in advance that is enabled by a true and final theory of reality) represents a basic misunderstanding of the epistemological relation between content and information. Information (relative information) is a retroactive abstraction of content that is performed by individuals, by virtue of their being (highly advanced) orienting systems. A condition for the existence of (relative) information, therefore, is the prior existence of content, which precedes it not contingently but as a matter of principle. So that the attempt to boil down content to information, however complex, necessarily fails to capture the relationship between the two as well as the most fundamental property of content: the fact that it is the result of a subject's effort to orient itself in a given environment, a process that precedes any attempt to conceptualize this activity as information.

Since we are headed toward a rather technical discussion, it will be useful this time, contrary to our usual practice, to start with a few basic working definitions. These will help us navigate the discussion about extensionalism, which in turn will allow us to comprehend the complex relations between information and content. Up until now, we discussed the reductionist ontology, which takes the world to consist of basic items, and holds that these are the only items that should be referred to (even if indirectly) in any scientific discussion. It is now time to talk about the question of this reference, i.e., about the language that the scientist should use according to the reductionist, whether the scientist does so consciously or not. After all, if reality consists of nothing more than basic items, then ultimately, the language of science should include only terms that in the final analysis designate and describe these basic items (and at most, maybe also various bundles or groups of these items). (Allow me to say again that I use the neutral term "item", and not other more familiar ontological terms such as "object" so as to avoid discussing their

N. Bar-Am, *In Search of a Simple Introduction to Communication*,
DOI 10.1007/978-3-319-25625-2_10

ontological nature, and in particular whether they are substances or not, and in what sense.) Therefore, reductionists have an utterly simple theory of meaning: they hold that the meaning of the terms used in the language of science is none other than their reference, or denotation. This reference is called "extension". Thus, the extension of a term is simply the item or items it extends to, or applies to. For example, the extension of the term "cat" is all cats, and the extension of "the morning star" is the planet Venus. Of course, strictly speaking, as we just noted, there are no cats and perhaps even no planets in the reductionist language but only different kinds of elementary particles (or other such items as the physicist will tell us in the far distant future); but for the sake of introducing the reductionist theory of meaning we can contend with using these more familiar objects.

A few more basic definitions. When a pair of terms denotes the same item or group of items, the terms are called "coextensive", implying that they have the same extension, or reference. For example, an "animal with kidneys" (known as a "renate" animal) is a term that denotes the very same group of items—the same animals—denoted by the term "cardate animal" (animals that have a heart). And the term "the evening star" denotes the same exact planet denoted by the term "the morning star": both refer to Venus. Now we are in a position to define extensionalsim. It is the attempt to describe our world with theories in which the meaning of the terms we use is simply their extension (i.e., theories in which the meaning of an expression is nothing more than the item or group of items it denotes). In an extensionalist theory—and this is the most important point for our purposes—coextensive expressions, and they alone, always carry the exact same meaning in all contexts (since meaning, as we saw, is ultimately nothing more than extension for the reductionist). In slightly more professional lingo we would say that a theory is extensional if it allows for the substitution of expressions (proper names, predicates, general terms and open sentences) only with coextensive expressions, and if wherever substitution of expressions with others is thus allowed, the new resulting whole is coextensive with the original one.

At first glance, extensionalism can seem like unnecessary pedantry. But this pedantry is very important, since throughout history it has emerged as a fundamental tenet of our understanding of scientific proof as well as of our understanding of the nature of scientific explanation.[1] Indeed, this is why we find it so hard to give up reductionism as the foundation of our scientific worldview: our conceptions of proof and of scientific explanation are bound up inextricably with extensionalism. Both conceptions are informed by the idea that a complete explanation is a proof, and that a proof consists in a series of substitutions of coextensive expressions, pursued until the desired identity statement is reached—an identity statement that all parties to the debate ultimately agree to be obvious. It was clear already to Descartes that the most basic and clear-cut statement of identity of meaning is that

[1]On the fascinating process of discovering that the extensional theory of meaning lies at the heart of logic, see my *Extensionalism: The Revolution in Logic* (Springer 2008). Readers who wish to delve deeper into the problems described in this chapter may also want to consult my essay "Extensionalism in Context", *Philosophy of the Social Sciences* 42(4, 2012), 543–560.

which describes an item's identity with itself (or a group of items' identity with itself)—and this, as we saw, is the basic extensionalist identity (a = a).[2]

To illustrate the notion of proof as a series of substitutions of expressions that have the same exact meaning, and also the reductionist's mistake in confusing reality with unattainable ideals, we can consider any true statement that comes to mind, provided that its truth seems to us absolutely trivial: "The morning star is the morning star", for example. Now imagine that we already have a theory in which the terms "the morning star" and "the evening star" have the same meaning (for example, because we are told, as a given, that they are coextensive within a reductionist framework, i.e., that they denote the same item, perhaps without our having previously noticed this fact). Then, on the truth of our trivial statement ("The morning star is the morning star") we could base the truth of a new statement ("The morning star is the evening star"), which perhaps in certain contexts did not initially strike us as trivial—for example, because we had not yet been exposed to the fact that this statement expresses a basic extensional identity, because we had not yet thought through the full logical implications of the definitions we received. This happens frequently in proofs: we discover that propositions whose status was initially not clear to us are propositions whose status is known and accounted for because they now clearly express an identity (or a non-identity, if we prove by *reductio ad absurdum*). And so, anyone who agrees with us that the first proposition is an undeniable truth, will agree with us, after the proof, that the new proposition (whose status was unclear at first) is an undeniable truth (or false, in *reductio ad absurdum*), and they may even regard it from now on as trivial.... Since from an extensionalist perspective, we changed absolutely nothing in the meaning of the proposition when we replaced one of its parts with another, coextensive part. Similarly, if we wanted to prove that 2 + 2 = 4, we could show that the expressions on either side of the equation are simply different names for the same group of items (and hence that the equation is reducible to 1 + 1 + 1 + 1 = 1 + 1 + 1 + 1). This, in fact, is how all valid scientific proofs are performed: the person who performs the proof gradually substitutes expressions and propositions recognized as having the same meaning (or in some cases as having a partially common meaning, but this does not affect the logic of the point we are making here), and in this way leads his colleagues to the understanding that what had seemed at first like a complicated proposition whose truth was unknown is now an obvious truth, as self-evident to any intelligent person as a = a. In other words, when we perform a proof, we reduce a proposition whose status is unknown to a self-evident formal identity that expresses a self-evident extensional relation: the identity between every item and itself, or any group of items and itself (or else to a no less self-evident false proposition, such as a ≠ a, as in the case of *reductio ad absurdum*).

[2]Sometimes, of course, we aim to prove not an actual identity claim but some other true claim—for example, about a proper containment between two groups of objects, like the one expressed in the proposition "All humans are mortal"; but this fact does not affect the logic of our point here.

Now, an extensionalist *theory* emerges when we accept a number of propositions as obvious basic truths, which we therefore call "axioms" or "definitions". Along with these axioms are articulated rules of inference that tell us how we are allowed to construct new propositions out of the basic propositions. It is important that these rules allow us to build new propositions out of the basic ones, but only by substituting coextensive expressions, as described above. In particular, our rules of inference disallow adding new expressions or constructing new propositions out of the axioms in any other way. In other words, the only sanctioned course of action is to take given propositions and replace them with coextensive propositions, or replace their parts with coextensive parts, and the result is always a new coextensive whole. A system of this sort is an extensionalist theory: it reflects content that has been arranged as an axiomatic deductive system, and thus prohibits—even at the syntactical level—the articulation of new propositions that would add content (from a theoretical point of view, not for a given user) to previously proved propositions, and ultimately to the axioms.

This is how things are from a theoretical perspective. And let us emphasize again that from any other perspective, and in particular from the point of view of our always-limited human condition, it is certainly possible that we learn something new from every new proof: after all, the set of all the propositions that are coextensive with a certain axiom (or any proposition for that matter) is never known to us in advance, precisely because we are just individuals getting to know their environment, and not any kind of omniscient God with unlimited powers of calculation (and even in the case of such a God, we know of certain limitations that apply to His ability to prove all true propositions—Gödel showed them in his famous theorem, which utterly destroyed the idea that the set of all the arithmetic truths is an extensionalist system, that is, it is impossible to derive it from any given axiomatic theory of arithmetic).

Reductionism, then, is at bottom essentialism, and like essentialism, it conflates reality and ideal: it tempts us to conflate the two even when we are not yet permitted to do so, i.e., before we have arrived at the end of days. Were it possible to inhabit a God-like extensional point of view, one that immediately identifies the content of an expression with its final and ultimate reference, and also immediately identifies all expressions that are coextensive with it, then from such an ideal point of view maybe no new content would be added to our world with every possible proof. But in all other scenarios, we may well have learned something new from having seen a new extensional proof (even if the very fact that it is an extensional proof implies that its conclusion should have been trivial to us, as it is nothing but a rephrasing of one of our axioms).

The extensionalist theory has two clear advantages over plain content conveyed between individuals. The first was presented above, so I will just recap it: if the axioms of the extensionalist theory are true propositions, then all the propositions capable of being articulated within the theory, all grammatical propositions within it, are also true. The system's extensional nature guarantees and clarifies this property, which of course characterizes all valid deductive systems. This is why extensionalism, which highlights this property, is central to our conception of scientific proof. For even if the axioms of our theory are not really known to be true but merely accepted by all parties, and even if they are accepted only for the sake of a given debate, we are guaranteed that all the propositions capable of being articulated within this theory will also be accepted by everyone, if only for the sake of the debate. This is because within such a system, any proper concatenation of propositions amounts to a complete substantiation of the last proposition in the system—i.e., it establishes the conclusion as coextensive with the premises. This first advantage, then, underlies the centrality of extensionalism to our scientific theory of explanation, which is based on the unity of explanation and of proof.

Reductionism is founded on the hope that one day all scientific knowledge will be expressed within a perfect extensional system: its terms will denote items that the last physicist will declare to be indeed the basic items of the universe (and maybe also certain select aspects of their presence in the world),[3] and all other truths will follow, almost of their own accord, from the extensional relations that the system dictates. Of course, as we have emphasized, any particular orienting system, any individual, may be aware or unaware that two expressions in our perfect theory are coextensive. But from a more abstract point of view (which presumes to be objective, and is basically the imaginary point of view of a God with unlimited powers of calculation), if the rules of our perfect theory allow us in principle to demonstrate that these expressions are coextensive, then these expressions have the same meaning whether or not anyone is aware of this, for they have always been coextensive, from this ideal point of view, and will always remain so (from the ideal point of view).

All this is good and well (even if in previous chapters we saw repeatedly that there is and can be no point of view that is also objective, and therefore that the expression "God's perspective" is an oxymoron). But we are not the sons of gods, and so we have to try and understand what the non-divine extensionalist point of view is—i.e., how content appears from an extensionalist point of view now, before we have reached those distant, ideal days when we shall possess the perfect theory and unlimited powers of calculation. Until then, the extensionalist knocks about, if we can put it that way, like every other orienting system, in the environment of his limited theory, and in the landscape of his limited knowledge regarding the theory's meaning.... In other words, he explores the particular and limited theory that he currently holds in search of new coextensions that are provable within it. He is restricted, of course, not only by his powers of calculation (which are capped as a matter of principle even if he is aided by the fastest computers), but also and especially by the fact that the axioms of his current theory, because they are not the final truths of the final perfect physical theory, express only the conjectured, limited knowledge currently at his disposal. This is the most pertinent point for our

[3]This point is of course extremely problematic, but I will not dwell on it here except to note that it is not obvious that the idea of a hierarchy between negligible and select relations is legitimate from a strictly reductionist extensionalist point of view, since it reflects a preference for certain relations over others, and this preference presupposes a point of view. The expert reader, however, will note that reductionism is thus essentialism and positivism conflated. It is thus irrevocably inconsistent.

purpose, since it reminds us that until that final ideal day, the extensionalist theory is arrived at always and necessarily through the abstraction of a given, limited content, and this content is of course obtained, for better or worse, from our limited human orientation in our environment. In other words, the content is prior, and necessarily so, to any one of its abstractions in the extensional framework. Thus, the scope of any extensionalist theory (which is not the final ideal theory) is limited precisely by the boundaries of the limited scope of the context-dependent speculations that are expressed in its axioms (as if it already were the final theory). This content is translated in the extensional theory into a set of permissions and prohibitions regarding the substitution of expressions—that is, into a set of syntactical rules. *"The extensional point of view", then, is expressed as the sum of these permissions and prohibitions—as the syntax of an extensional system—but without the further context that gave them their meaning in the first place and by virtue of which they were formulated.* In short: syntax is content that has been abstracted. And within extensional systems in particular it is content whose context has been, so to speak, stripped off.

To understand why this seemingly pedantic point is so important and interesting, and how it shows us the limitations of our scientific worldview insofar as this view is reductionist, it will help to consider the following well-known example. Recall the true proposition that "The morning star is the evening star". It is a fact of history that the discovery of its truth (the discovery that the planet visible in the sky before sunrise and therefore named the "morning star" is the same one visible after sunset and therefore named the "evening star") is attributed to the Babylonians. The Babylonian astronomers made a spectacular discovery by calculating the motions of what had previously been thought to be two celestial objects and discovering that they amounted to a single orbit, leading to the assumption that the motion was in fact that of a single celestial object. Now let us consider the Sumerians, who preceded the Babylonians and, as far as we know, still thought that the morning star and the evening star are distinct objects. And let us ask ourselves a simple question: *How can the reductionist scientist discuss a wonderful discovery of this sort from an extensional perspective?* Could the reductionist discuss it even if he possessed the God-like, omniscient extensional perspective?

Both the critics of extensionalism and its most ardent defenders agree that a consistent extensional description of this kind of discovery encounters fundamental difficulties, difficulties whose theoretical core the reductionist cannot resolve without abandoning (momentarily but unavoidably) his reductionism. These difficulties are resolvable, but only in retrospect and in a necessarily local manner, and never in principle: they can be circumvented only if the extensionalist agrees for a brief moment to betray his extensional point of view and acknowledge that the content of a term is more than the item or group of items it denotes, and that context is more than a set of permissions and prohibitions—presupposed and unsupported by further reasoning—regarding the substitution of terms. The significance of this pedantic point for our purposes is great. It means that contents precede information as a matter of principle, and therefore that the most fundamental tenet of reductionism—the idea of a complete reduction of content to information—is based on misapprehension.

To witness this, let us proceed with our example. Suppose for a moment that we have succeeded in expressing Babylonian science as a deductive theory (a task far more complex than typically assumed, but we will not dwell on that), and that we now possess an extensional theory that expresses Babylonian knowledge in its entirety. In such a theory, the identity "the evening star = the morning star" is of course already given, so that even an individual of very average intelligence, to say nothing of a divine user, can easily reduce this proposition to the trivial equation "the morning star = the morning star." This is unsurprising, since the expressions are already known to the Babylonian scientist as different names of the same object: they are given to him as coextensive and appear as such in the most basic definitions of the reference of the terms of his extensional theory. Yet recall that the problem before us is not whether the Babylonian discovery is true, or how to prove its truth, but rather how can an extensional theory discuss the Babylonian discovery; and in particular, how can it discuss it *as a discovery*?

"The morning star = the morning star" is a truth that no one would ever consider denying. By contrast, "the evening star = the morning star" is a proposition that the very sensible Sumerians had taken to be simply false, from which it follows that reducing it to the trivial truth "the morning star = the morning star" is simply not possible. From any extensional perspective we choose, whether human or divine, then, the Babylonian discovery is either a self-evident truth, requiring elementary powers of deduction, or a plainly false absurdity, the kind that can never be shown to be true in our theory. Thus, *the discovery-qua-discovery disappears*: it has no place in the extensional point of view.

To underscore this important point, consider a Sumerian astronomer. Let us call him "Utnapishtim". Now consider the proposition (true, of course): "Utnapishtim did not know that the morning star is the evening star". If in our extensional theory the terms "morning star" and "evening star" are coextensive (as they should be from the Babylonian extensional point of view), then from the same point of view we are committed also to the utterly ridiculous claim that "Utnapishtim did not know that the morning star is the morning star".[4] This is the absurdity embodied in the reductionist attempt to ignore or explain away context. It is the absurdity embodied in the assumption that meaning is really nothing more than denotation. This assumption yields a picture of the world in which systems exploring their environment "discover" in it only trivial aspects that they already knew in advance. It ignores the fact that orienting ourselves in our environment involves constant and

[4]This example appears in the literature in multiple versions. Willard Van Orman Quine, for example, the most important proponent of extensionalism in the twentieth century, discusses the example and its local solutions in many of his essays. Quine recognizes, of course, that the solution to the extensional challenge is necessarily local—i.e., that a general or principled solution to this problem is impossible. Indeed, as I show (Bar-Am 2012), the recognition of this basic fact is the very heart of his theory and the source of all his varied and often seemingly eccentric philosophical positions. However, in this as in many other positions, Quine follows Russell, who first noted the problem in his famous "On Denoting" (Russell 1905), and Frege, who discussed it extensively in his essay "On Sense and Reference". For historical context and a discussion of the limitations of the solutions to this problem, see my paper ibid.

subtle changes in our speculative representations of that environment, and therefore that it is at once a task and a meta-task (a state of affairs that is utterly unacceptable from an extensional point of view). It ignores the fact that orientation is an extremely unique task: even when its goals are more or less fixed, they are never more than that (they are never set in stone).

We have arrived at the second important advantage of the extensional theory, which will help us broaden our conclusions regarding the limitations of reductionism. And with that we will conclude. The discussion that follows is in large part a repetition of the discussion above, but its importance for us is great, for it also extends our conclusions to the world of "artificial intelligence", the world of computers and their "communications".

The second prominent advantage of the extensional theory, then, is that it can easily be expressed as a fully formal theory, which in turn helps us better understand what we are seeking when we seek to express content as (relative) information. In mathematics and logic it is common practice to strip the theories almost entirely of their meanings, even their extensional meaning, by deliberately ignoring even the fact that their expressions denote particular items or groups of items, and that their original propositions were an attempt to give deductive form to a certain body of knowledge. This ignoring, or stripping away, leaves us with the theory's *logical skeleton* (which reflects, as noted earlier, a highly abstract and purified aspect of the original context) and it is therefore described in the literature as a "formal system". Instead of a theory concerned with a world of basic items and certain permitted basic interactions among them, we now have only expressions that we treat as almost entirely free of content, i.e., almost as mere forms, along with permitted and prohibited interactions among these empty forms. This is the purest form of relative information that was mentioned in the previous chapter: the formal system expresses only a very abstract, very lean aspect of the original content relations, and this pared-down version obviates any talk about how the original terms gained their original meaning and how the original propositions arrived at their particular arrangement (namely, according to their content within a particular context).[5] Information, then, is content that has been stripped of many of the aspects that made it content in the first place[6]: first, we gave the terms of our scientific

[5]In the literature, the formal system is often described mistakenly also as a "formal language", but it is important to note that it is not a language at all but rather an abstraction of content arranged deductively (which can, of course, also be expressed *in* a language). As opposed to an axiomatic formal system, *in a language both true and false propositions are expressible*. Not so in the axiomatic theory, formal or extensional. In fact, in an important sense, which I cannot say much about within the scope of this discussion, *reductionism is founded on this blurring of the distinction between a theory and a language, and so between a semantic and a syntactic description of content*. For the sake of expert readers I will merely point out that a reduction of the semantic to the syntactic requires a meta-language that the strict reductionist cannot allow.

[6]It is perhaps worth emphasizing that formalization is one of several ways to pare down content—i.e., to transform it into relative information in the sense described above. Any simple kind of disregard (or ignorance) turns a given content into relative information of this sort: if, for example, I copy a Japanese book word-for-word without understanding Japanese, I am conveying to

theory clearly defined extensional senses and arranged them as a deductive system, and then we further stripped the newly established extensional theory that we received of any hint of extensionality, transforming it into a fully formal system. This is how content is pared down to information in ordinary scientific practice.

Note that I say that the original theory was "pared down" and not "reduced"—because neither the extensional system nor the formal system *preserve* the richness of the original contents, their rich meaning for the scientist and user. Thus they are not successful comprehensive reductions of these contents. However, they express attempts to isolate and reflect certain aspects of the contents that were present in the original context and indeed to facilitate a more efficient understanding of their implications, and this is why this information is called "relative": it was obtained in relation to the context in which it emerged, it is relative to the body of knowledge whose pared-down and highly abstract aspect this information now expresses.

And yet, note that any computer program can be described as such a formal system of relative information. In other words, every action by a computer and every computer is an illustration of the potential embodied in a formal system. This means that our present discussion of extensionalism, which has focused on the limitations of the reductionist model of scientific explanation, can be reproduced with only minor changes as a theoretical discussion of the limitations of artificial intelligence. This fact is important for our purpose: our discussion of extensionalism has effectively included a theoretical summary of the debate about the limitations of artificial intelligence, and that is no small feat.

To grasp the significance of this, imagine for a moment that Sumerian science were expressed as an extensional theory and then converted into a computer program through the steps described above. Now, the Sumerians, recall, mistakenly thought that the proposition "the morning star is the evening star" is false (just like every other proposition of the basic form "a = b"). Suppose, then, that a Sumerian computer meets its colleague, a Babylonian computer, at a conference about the planet Venus. How do they communicate? What might they say to one another at the symposium on the question of whether the evening star is the morning star? Are computers capable of considering open questions intelligently? Can they "realize" that they are all dealing with the same problem even when their operating systems are different and the conditions for solving the problem have not been defined for them in advance in precisely the same terms? How? At our imaginary conference, the Babylonian computer steps up to the podium and proves that the morning star is of course the evening star. Given how the computer has been programmed, this is a simple one-step proof. The Babylonian computer "sees" no problem here. The proof is trivial. Now the Sumerian computer goes to the podium and proves the opposite—again, as an obvious truth. It also "sees" no problem. And there is no

(Footnote 6 continued)

Japanese speakers the content expressed in the book as relative information. Similarly, the copying of genetic content, as this process occurs in nature, is its copying as information. This highlights the fact that the term "absolute content", just like "absolute information", designates an ideal, an extreme case that is unattainable in principle.

meaningful controversy between them! They do not even "speak" the same "language" (in other words, they operate according to different programs, which are, as we saw, pared-down expressions of contexts that the programmer has rigidified into lists of permitted and prohibited substitutions of expressions). In short, the two computers do not communicate.

We could of course try to unify the two programs. But the important point is that a simple merging of these programs yields a program with an overt contradiction. So that any other, more sophisticated solution of the problem of how to unify the programs—for example, giving the computers localized instructions on how to handle such contradictions or how to handle this particular contradiction—requires the programmer once again to *consider the contexts* within which the contradiction may become harmful or undesirable, to make conjectures about future problems that may occur as a result of it, and such actions by necessity involve peering beyond the formal description of the context toward a future reality as the programmer imagines it, looking to his private environment as a programmer with certain goals, to his reasons for building this computer and writing its program, and so on.... And yet, this kind of perspective and these kinds of considerations are precisely that which the consistent reductionist perspective, whether extensional or formal, does not permit.

What happened here? The fact that the notion of discovery does not sit well with the idea of a perfect theory (extensional or otherwise) should not surprise us: obviously, those who know everything do not discover anything. But our examples reveal much more than this. They illustrate that anyone who observes things from an extensional perspective of one sort or another, let along from a formal perspective, cannot even understand what a discovery is, no matter how great are his computational abilities: the discovery-qua-discovery remains outside the bounds of every consistent extensional or formal description of the course of events, precisely because the discovery-qua-discovery is a movement from one context to another while preserving a purpose (the relative constancy of the purpose is that which grants the transition its intelligibility). This cannot occur in extensional and formal systems. Since there is no room for purposes or contexts in the reductionist picture of the world, there is no room for them in the extensional or formal world. Thus, it is possible to build a computer that behaves as if it has a purpose, but a computer cannot actually have one. Only its designers can.

Whenever any one of the entities who orient themselves in our world learns something new, or even just adjusts itself to its environment, whenever any accidental or deliberate communication passes between us—a "discovery" of the kind discussed above has been made, a purposeful moment has occurred. Orientation is a constant movement between contexts that preserves a certain constancy (not rigidly defined) of purposes (and even the purposes themselves in some cases tolerate a degree of flexibility).

At this point, some of my students or the computer experts among my readers may feel a rising sense of disapproval. They would like to draw my attention to the incredible wealth of techniques that software engineers and computability researchers nowadays use to conceal (to overcome?) the straightforward contradiction that rears its head every time a computer has to update its operating system—that is, every time it has to imitate what we typically call "learning". I want to emphasize to such readers that I am well aware of the sophistication of these solutions. But allow me to draw your attention to two points. First, no one doubts that computers can simulate for us processes that they are not actually undergoing. But strictly speaking, the fact that we regard our computer as the same machine before and after the software update is grounded precisely in the fact that *we* observe it from the outside, as creatures who possess purposes (and ascribe such a purpose to "it") and are therefore willing to blur the distinction between two different objects: we carve our environment out of reality by virtue of our goal-directed existence, and therefore unite two objects that strictly speaking are not one and the same into an apparent single object. Ironically, it is the strict reductionist who really ought to be the first to jump up and point out to us that this move is not legitimate in principle—that such inaccuracies should not be tolerated, that vagueness is the enemy of science and that therefore every software update sets before us a new computer. It is thus the reductionist, first and foremost, who cannot admit that a computer undergoes a process of learning….

Second, in the final analysis, every OS update performed by a computer is always *local and retroactive*. Ultimately, it reflects the aims and purposes of its users. These are two weaknesses that a consistent reductionist cannot afford to accept. The world of artificial intelligence is expanding in recent years in truly incredible ways and I do not doubt that computers will be able in the near future to encompass and replace almost every aspect of human activity with a satisfactory level of success. But so long as they do this in more or less the same way as today, i.e., through the expansion of techniques that retroactively simulate learning and intelligent purposeful movement between contexts, the rules of the extensional/formal game cannot change, and so our conclusions must remain the same. For the common basic element of all of these simulation techniques is a sleight of hand: an attempt to predict in advance, through various means, limited aspects of the behavior of computer users. As such, computers who simulate intelligent conversation are merely expressions of human theories of behavior, their operating systems "predict" the computer user's behavior and "react" accordingly: the better the theory, the more complete is the illusion that an intelligent entity has reacted to the user's communication attempts, the more complete the illusion that actual communication has taken place. But as we have already stressed, all strict theories of behavior are flawed in principle simply by virtue of being sealed and final. Thus, simulation of orientation is not a case of orientation but only of an imitation of orientation within a setting whose boundaries are stipulated in advance, and in any way are pre-defined. Ultimately, then, we do away with the discovery-qua-discovery by writing it in advance into the OS. And that, of course, is no discovery at all. In this subtle and important sense, computers do not know their way around their environments at all: they just do a good job of imitating such activity. Whoever wishes truly to reduce content to information seeks in effect to deny the existence of contexts. In doing so, he loses the context that set him on his scientific query to begin with, he loses the meaningfulness of his scientific problems as well as the meaningfulness of his theories and tools of proof.

If you find yourself somewhat worn out by this lengthy discussion, here is its gist. Extensional and formal systems do not allow for a successful discussion of discoveries. But the word "discovery" does not represent some sort of singular, mysterious case whose enigmatic nature challenges the reductionist picture of the world. It functions as a marker of a much more general problem: it is but one more aspect of the general problem of emergence that all communication scholars face: the existence of self-orientation and of communication (and of course, of consciousness). This is the problem that is stuck like a bone in the throat of reductionism. For an environment is a flexible set of possible meaningful behaviors and their often unpredictable results in light of more or less (but not fully) fixed goals, and we do not know how a consistent reductionist can recognize the existence of such goals or such a set of meaningful behaviors. Every action by every orienting system in its environment reshapes it, if merely slightly, as well as its environment and so some of its goals in ways that are often rather predictable but sometimes cannot be accurately foreseen. It follows that orientation is an action whose results are somewhat open as a matter of principle, and this openness is not adequately expressed in the reductionist descriptive frameworks (for it is meaningful openness, not randomness). Because we are individuals who are surrounded by an environment, we do not know what tomorrow will bring, nor even who exactly we ourselves will be, and this is why we need communication: to coordinate as best we can a change in ourselves and in our environment in light of our goals, which are also in gentle flux. This is why the study of communication is such a threat to reductionism, and why it is so difficult.

To underscore the general meaning of our discussion let us return once more to the example of our Babylonian astronomer Utnapishtim. We saw that the proposition "Utnapishtim discovered that the morning star is the evening star" is problematic from an extensional point of view (because it can lead us to the ridiculous claim that "Utnapishtim discovered that the morning star is the morning star", or to the ridiculous claim that Utnapishtim proved a contradiction). But the propositions "Utnapishtim felt/dreamt/guessed/knew/believed/was excited/was happy when he realized that…" or even just "told his wife that…the morning star is the evening star" *all pose exactly the same problem*—any extensional description of these facts is necessarily limited in the sense described above because it is bound up with the inevitable limitations that apply to the reduction of content to information (or more precisely the paring down of content to information). Our world, so it seems, can never be fully accounted for in terms that fully ignore our existence as orienting systems with an environment and (therefore) with goals.

Reductionism is an all-out war on contexts. The cause of this battle is just: every context does indeed represent a case of vagueness, an ambiguity that from a more rigorous perspective ought to be cleared up. The difficulty lies in the fact that every local reductionist victory over a given context is enabled by a new context. The reductionist is right to point out that this kind of move often represents progress in our scientific knowledge. But he is wrong when he claims that we can attain an understanding of our environment and an expression of this understanding in language so perfect as to render unnecessary and indeed impossible any further assessment of the reference of the terms we use. Moreover, for a perfect speaker of this ideal language/system, communication and orientation are undoubtedly an illusion. In all other cases—i.e., until the doubtful realization of this ideal—the existence of contexts is built into our existence as orienting Systems.

Since scholars of communication take the possibility of orientation and communication as the most obvious presupposition of their field of research, they repeatedly expose the limitations of the latest reductionist picture of the world. As we will see later on, this constitutes the greatest contribution of the scholar of communication to the scientific effort, and it is why the status of communication studies is so loaded, and so crucial.

Bibliography

Bar-Am, N. (2008). *Extensionalism: The revolution in logic*. Dordrecht: Springer.
Bar-Am, N. (2012). Extensionalism in context. *Philosophy of the Social Sciences, 42*(4), 543–560.
Linsky, L. (1967). *Referring*. New York: Humanities Press.
Quine, W. V. O. (2008) Confessions of a confirmed extensionalist. Cambridge: Harvard University Press
Russell, B. (1905). On denoting. *Mind, 14*(56), 479–493.

Chapter 11
A Note on the Intelligence of Computers

Can computers think? Debates regarding this confused question fill the pages of the popular press and especially of that of the pseudo-scientific kind. It seems that with every technological advance achieved by researchers of artificial intelligence we are once again assured that we are just around the corner from witnessing computers who think and act like human beings, and of course surpass them in many respects. Prophecies of doom and of glory about the frightening or inspiring day when computers take over the world sell newspapers and movie tickets, and somewhat surprisingly, also sustain the careers of some dramatic academics and researchers. Though we will return to this question later on from various angles, I think we have by now accumulated enough tools to realize that, strictly speaking, these debates almost always involve a confusion between (at least) two distinct questions. The first is a question that presumes to be context-free, i.e., to be essentialist—namely: Can human thought, that is, the capacity of humans to think, be faithfully captured and represented by means of a single computer program? From reading earlier chapters, you understand by now that the answer to this question is clearly negative, simply because it is not possible—as a matter of principle—to capture by means of a single deductive theory (formal and/or extensional) the infinite systemic possibilities of any single slice of reality, let alone a conscious human mind, whose awareness to that theory may influence its future patterns of behavior.

The other question is of a very different sort: Is it possible to represent by means of a computer program a certain limited aspect of human behavior in a given and well-defined context (or cluster of contexts), when this aspect has been carefully arranged in advance as a deductive theory and stripped of all of its possible expansions (or else—and this amounts to the same thing—whose permissible expansions are carefully defined in advance), in the manner described in the previous chapter? This latter question is posed without any regard for the question of what human thought ultimately *is*, as well as for the question of whether human conduct, including all forms of human creativity, can be arranged and predicted by deductive means, and so it should be clear to anyone who accepts the gist of what I said in the previous chapter that the answer to the second question is positive.

For the sake of illustration, consider if you will a human actor who does a good imitation of some other person in a certain situation. It may well be the case that although the imitation is very good, the actor does not fully understand this person's

N. Bar-Am, *In Search of a Simple Introduction to Communication*,
DOI 10.1007/978-3-319-25625-2_11

motivations, and in any case, he certainly can neither identify with these motivations in every imaginable circumstance nor predict them in every possible context. In other words, he cannot know how the person he is imitating so well in a limited context will act in a new and unpredictable context, how he will respond to an unfamiliar situation (even the person himself does not know this). The actor does not need to have an opinion on this matter in order to imitate the person well in a limited context—for example, in a series of pre-scripted scenes. *Just as Descartes could not divine the overall truth about the totality of mechanisms in which a given reality slice can potentially partake, so the actor cannot guess the sum of all contexts in which the object of his imitation can in principle partake as part of a broader system, and thus cannot guess his future behavior, which may be new and original, and in any case is original from the point of view of anyone who had not anticipated this particular context. The same is true of the computer: however fast it may be, being nothing more than a formal system in material form, it can never perfectly predict the behavior of a person in the countless contexts in which that person may be placed.* This is true regardless of the fact that in the near future the operations of computers will indeed reflect very extensive knowledge about people's ordinary activities in the vast majority of familiar contexts. It is important to understand that the future speed of computers or the amount of information they are able to contain has absolutely no bearing on this problem, and that those who predict that a radical improvement on these fronts will someday make the problem go away, simply do not understand what a computer is and what the problem that we are discussing here amounts to. In some rare cases, my own impression is that they know full well the difference between computers and human beings, and are simply not being candid.

And another point. An actor may imitate someone else well without himself knowing *how or why* he is able to do so. This is because mechanisms that are very dissimilar from a structural point of view can function in an identical manner in a given context, or in other words, because *understanding how a mechanism works is a different matter altogether from successfully imitating its operation in a particular delineated context*. I think this is a key fact, and one that aptly describes the state of things also in the case of a computer program that simulates human behavior in a given context or a limited cluster of contexts. Such a program is not necessarily, and probably not at all, an accurate description of the actual dynamic that underlies the human behavior it simulates (it is easy enough, after all, to build two—or more—different computers, or computer programs, that perform the same action equally successfully; first-year students of computer science regularly do this as an exercise). It is clear, then, that much the same applies to human behavior or to the activity of the human mind as an object of research: a scientist can (with a computer program, for example) successfully simulate a certain aspect of human intelligence that she wishes to study and which she has already arranged as an extensional theory (such as, for example, human behavioral patterns given a certain limited environment as the various behaviorists tried to do, and even Pavlov and his disciples before them—all of them following in the footsteps of Descartes, who viewed all animals, with the possible exception of human beings, as machines). But this scientist would still be very far from understanding or even just successfully representing the possible behaviors of her subjects in countless other possible contexts—that is to say, she would be far from answering the general, theoretical question of what influences the behavior of humans (or of dogs) in the countless other contexts in which they can potentially be placed. Mario Bunge aptly described this basic error—of identifying a thing with the sum total of its functions in a certain given context—and fittingly named it "the functionalist fallacy" (i.e., the suggestion that once you have imitated the behavior of your research subject (in a limited context) you thereby also understand "what makes it tick", to use his own catch phrase). It is only fair to note that this notion is Bunge's way of re-articulating an argument that had already been given a very clear formulation by Michael Polanyi.[1] And we have seen already that this is the central flaw of the reductionist picture of the world: it is a modern incarnation of Descartes' misleading rule of method. It is this mistake that underlies the endless grandiose announcements about conscious computers and similarly ridiculous promises to "back up" our human brains (as if we were bags of memories, or maybe even bags of memories deductively arranged), as well as talk about the not-too-distant future, alternately utopian and dystopian, when computers will replace us as scientists or as romantic mates.

I want to stress that I have no intention to deny that computers already perform almost fantastical tasks or that they will aid us in accomplishing unimaginably incredible feats in the very near future. I am no less excited (and maybe even more) than certain popular eccentric gurus of technological innovation about the possibilities embodied in the technological revolution, which is still in its infancy and will no doubt profoundly alter our nature and our environment. Nor do I intend to detract from the immense value that we gain from an extensional or formal study of one or another aspect of the human body, and even of what we call—misleadingly, and out of habit and tradition—the human "soul", "psyche", or "mental life"; that is to say, the value that we gain from studying the mind by likening it to some kind of mechanical mechanism or computer program. What I am claiming, rather, is that every metaphor obviously has advantages and disadvantages, and that it is precisely the constant expansion and amplification of the metaphor of man as a machine, or of mind as computer program that helps us identify more accurately and highlight

[1]The functionalist fallacy underlies many different reductionist tendencies that we will not address here, including, most prominently, the psychological school of behaviorism. Behaviorists advise us to abandon the study of intentions and their various inner representations and focus instead on the study of predictable behavioral regularities. Of course, without any understanding of intentions, behaviorists find themselves in a predicament similar to that of the actor who imitates someone's behavior accurately in one context but does not know how to extend the imitation into another context. For the only thing that leads from regularities in a given context to their possible and unknown elaborations in other contexts is intention. As far as I am aware, Polanyi (1952) was the first to express this insight as a devastating criticism of the argument that computers can think. This was a particular application of his broader argument against the reduction of machines in general to their purely physical descriptions (Polanyi 1951, p. 21). Incidentally, Polanyi was a friend and colleague of Alan Turing, whose work is discussed below. Polanyi's essay on cybernetics was written explicitly as a critique of the idea that a Turing machine succeeds in capturing something of the nature of human thought.

the limits of this metaphor: every advancement in our knowledge about what computers are capable of doing sharpens our understanding of what computers cannot do as a matter of principle—i.e., what they will never be able to do so long as our understanding of the limitations of formal or extensional systems does not undergo a radical transformation.

The functionalist fallacy lies at the heart of the ridiculous test known as the Turing Test—a pair of words that the AI public relations machine returns to like a magic mantra. Turing, if you do not know, was a brilliant and original mathematician who began his career with an outstanding reductionist breakthrough: he proposed a simple model that represents one limited aspect of our intellectual capacity within a well-defined context: he reduced the human ability for arithmetic calculations (our ability to perform addition and multiplication) with the aid of an abstract model that is now known as the "Turing machine" and which helps us understand, from a very general point of view, what computers can and cannot compute. This was an unprecedented and constitutive contribution to the budding field of computer science. Later, Turing tried to explore a possible generalization of his achievement in response to the confused and contextless question (in other words, the essentialist question) of whether computers can think like humans. Because this question is clearly embroiled in the functionalist fallacy, the answer to it is decidedly negative. Turing undoubtedly realized this more lucidly than all those who surrounded him. But he tried to remove the major ambiguity that is inherent in his new task by minimizing it in a way that has confused and misled many of his admirers, philosophers and engineers alike: he claimed that on the day that a computer succeeds in "fooling" an ordinary person into believing that it thinks like a human being, we will be justified in declaring that computers can think…. This logic is analogous to saying that when an actor succeeds in tricking someone into believing that he is in pain (or having an orgasm), we can declare that he is actually in pain (or having an orgasm). And it attests above all to the desperate state of the reductionists, who started out as sincere and uncompromising seekers of truth and end up making do with a successful deception of others as well as of themselves in an attempt to hold on to their false picture of the world at all cost. Those among them who are serious and sober know that their public relations are based on a falsehood, but they are nonetheless prepared to regard this PR as a pretty close representation of a final truth. And this is precisely the distinction that is gradually becoming clear between the two very different tasks of understanding and imitation; it is the same old confusion between the ability to simulate a phenomenon in a given, limited context (or a given set of such limited contexts), and the ability to understand and embody it.

Agassi (1988) summed up this discussion with a simple and very useful clarification. Turing, he noted, tried to ignore the fundamental fact that his experiment is not final as a matter of principle. In this sense, it is more like an open heuristic guideline than a real claim about the ultimate nature of computers or the essence of human intelligence. And this is the very same fact that artificial intelligence public relations repeatedly obscure. After all, a machine that simulates human behavior in a given context is no more than a hard-wired theoretical expression of our speculation that human behavior in that context is predetermined and subject strictly to a system of deduction…. In other words, *a computer program that simulates behavior is a deductive representation of a theory of behavior.* And yet, we have already noted that the very idea of a final and comprehensive theory of human behavior is highly implausible, and not merely because awareness to the fact that our behavior is predetermined may lead us to alter this behavior in the next moment/context. In the same way, being "fooled" once by a computer in a Turing test can lead us to figure out how it is that we have been fooled and to "call the computer's bluff" the next time it attempts to do the same. In turn, of course, the deceiving program can be improved and taught new "tricks", and so on ad infinitum. Similarly, the consistent failure of computers to simulate human intelligence successfully is not proof of their inability in principle to do so sometime in the future. Thus, if and when we mistakenly treat the Turing test as a metaphysical test —that is, a test designed to determine the fundamental nature of mind or the essential ability of computers—the discussion is easily sidetracked into a hopeless series of meta-debates. Ask yourselves, for instance, whether a computer can program a computer that will pass the Turing test; and if so, whether a computer can then program another computer that will successfully call the first computer's bluff; and so on…. What all of these debates and meta-debates obfuscate, argues Agassi, is the simple fact that the challenge is open and never ending because computers are no more than simulation devices: they help us simulate the possibility (which we know in advance to be false) of providing a complete and comprehensive description of the behavior of humans in their environment, as if they were deductive systems, as well as the (utterly false) possibility that a complete and comprehensive description of all contexts can in principle be produced.

Context, as we saw, is enabled by orientation challenges, or goals. In light of the goal the distinction between different stimuli is blurred so that they are perceived as instances of the same "thing", and thus as meriting the same response pattern. This is how (private and inter-subjective) environment is created out of (objective) reality. Human behavior is a constant movement between contexts, between environments, as well as a constant critical assessment of that movement according to various new orientation challenges. It is only in the context of such challenges, or goals, that this movement, which from an extensional point of view is utterly meaningless, carries any meaning. It seems, therefore, that we cannot get rid of the bathwater without also throwing out the baby because the water and the baby are two aspects of one and the same thing.

Let me clarify again that I have no doubt that the ability of computers to "pass" as human has fascinating and important implications for the engineer, the marketer, and in some cases also the scientist. There is no doubt, for example, that our legal system has to adapt itself, and quickly, to the many possibilities embodied in computer programs whose operations coincide with the conduct of human beings in many contexts. They can easily be mistaken as making decisions when in fact their actions have been programmed into them. I also have no doubt that with delicate human guidance (and maybe later also with the guidance of computers), computers will be able to perform tasks that no one can yet even imagine. And however much or little computers improve in their ability to simulate human intelligence, there is no doubt that humans are very poor at imitating certain aspects of the activity of computes, most notably the speed with which computers compute, just as computers are poor at imitating our ability to re-examine contexts in which we previously reached a dead end.[2] The more important and interesting point is that the cooperation between humans and computers, in almost all interesting contexts, is in my view strikingly more intelligent than the actions of either one of them alone—and it is so precisely because of the marked difference between humans and computers. For example, a computer that relies on image recognition and descriptions that humans have supplied it with can gradually attain a reasonably good ability to give new images an appropriate human title. This, of course, has nothing to do with the computer's ability to perceive and understand but rather with its ability to mediate images to the human titles. And if its database someday grows to contain all the titles that humans have ever given to images, it may well be able to make pretty good predictions about the next title we would give a certain image. But from a theoretical point of view, these incredible facts are of little significance: from a theoretical perspective, saying that such a computer "sees" or "perceives" the images is no different than saying that it thinks or that it is in love; it is just another incarnation of Descartes' fundamental error.

I end this brief note with an anecdote, which may not add new substance to the discussion but does demonstrate how well aware those at the forefront of developments in the field of AI are of everything that has been said here, in contrast to the false promises of those who spearhead the public relations of this field—which really has no need for their populist protection (and in particular for their claim that computers think and feel or that the mind is just a computer). At a conference of logicians I attended several years ago, I met a gentle soul. This man approached me with timid curiosity after overhearing someone talk to me about my book on extensionalism, and asked me politely what the term meant. I gave him a short,

[2]See Popper (1950a, b), and especially the final paragraphs of part II, for a fascinating attempt to explain this point. The thought experiments proposed there, which concern the limits on the ability of predicting machines to predict their own future states, are vital background for Agassi's outstanding examples of possible elaborations of Turing's test (Turing machines teaching other Turing machines to identify Turing machines that impersonate humans, etc.,). A computer, Popper argues there, given enough time, can of course prove almost countless arithmetic truths, but it cannot tell the interesting ones from the uninteresting ones. Given, for example, that a computer has shown that $2 + 2 = 4$, it will also prove that $2 + 2$ is different from $4 + 1$, different from $4 + 2$, different from $4 + 3$, and so on. Of course, we can try to program a computer to distinguish between interesting and uninteresting truths using an algorithm that expresses an assumption about how this might be done, but this does not solve the problem—it merely pushes the problem back to another meta-level, since our assumptions about what is deserving of interest change according to context and require context in order to have any sort of grounding. Suppose, Popper adds, that we have absent-mindedly entered inconsistent axioms into a computing machine. It will proceed to derive from these axioms almost countless valid theorems without realizing the utter futility of the task. But how, asks Popper, would this computing machine handle the context-sensitive question of how to fix the faulty system of axioms? And how would it distinguish between interesting and uninteresting solutions to this problem?

three-sentence explanation, and when I saw that he required no more than that, I knew that he must be a mathematician. Then he introduced himself, and I realized that he indeed did not need me (nor even Quine, the father of the term "extensionalism" and one of the twentieth-century's greatest philosophers) to understand the meaning of extensionalism and of formal systems, or the ultimate limits of computers as tools for simulating thought. It turns out that the man who asked me to explain extensionalism is one of the world's leading scholars of computability and head of the Microsoft research and development team whose work centers on cloud computing. Like Molière's protagonist who discovers to his delight that he has been speaking prose all his life, my interlocutor must have noted to himself that extensionalism is one more label for a field whose details he has in any case already been studying, regardless of their title, and that that was good enough. The next day, with unexpected generosity, he came to hear my talk at the conference, and then invited me to join him at a nearby café for a conversation with another researcher, also a mathematician and a leading world scholar of theoretical computer science. We sat together in the beautiful streets of Nancy, where the conference was held, and sipped coffee in a picturesque café overlooking a glorious cathedral several hundred years old. We talked about machine translation, a topic in which all three of us shared an interest. Since the discussion did not involve fine technical detail, I could participate with ease and pleasure. My interlocutors wondered whether Quine had contributed in any way to the debate about the limitations of machine translation. I answered that Quine had not made any substantive contribution to this question beyond clarifying several fundamental points that were later elucidated very nicely by the Israeli mathematician and philosopher (Bar-Hillel 1964). Quine, I said, had merely made the obvious observation that computers would never be able truly to converse like humans, and thus would never be able to translate like humans, because of their structural limitation as extensional systems. My interlocutor rolled his eyes in the manner of someone who keeps being told something that he has known full-well from early childhood. He smiled, and told me the old Russian tale of Vanya and Grisha who lose their way in the forest. Suddenly, the tale goes, they see a bear charging at them. They start running for their lives, but the bear is faster. Still running and panting, Grisha turns to Vanya and says: "The bear is faster than both of us, Vanya, we can't outrun it. Let's just stop… we know how all of this will end". "Of course the bear is faster than us, Grisha", Vanya replies. "But I'm not trying to outrun the bear. I'm trying to outrun you…".

The moral is clear. What the good man was trying to tell me is that, from a theoretical point of view, Quine and Bar-Hillel are obviously right: computers simply cannot think, and clearly will never talk *exactly* like humans. But this point of view is not particularly interesting for the engineer, the computer scientist, the researcher of machine translation. Engineers do not ask themselves whether the computer can think or translate exactly like a human being (whether man can outrun the bear), but simply how to create a program that will replace one or another practical aspect of human intelligence within a given budget (that is, how to outrun their human peers in a given context). The question that concerns engineers, he wanted me to understand, is not whether future translation machines will converse and translate just like human beings, but rather when, in light of current developments in computability, and in which sub-field, will humans start paying software engineers to replace the human translator with their translation machines; in other words, at what point and in what context will computers replace the human translator *despite their known structural limitations*. Now it was my turn to smile and roll my eyes, since my interlocutor had just effectively articulated the Turing Test. And he was right to do so, of course, because in what concerns the practical domain, the Turing Test is indeed the only test that ought to interest engineers; but the misleading aspect of Turing's insight has to do with the fact that he allowed people to conclude, mistakenly, that the engineer, in doing so, also solves the general philosophical problem. I was very pleased that we had "reached" this understanding as a mere trivial clarification. And pleased in particular to see that this point appeared to be in obvious agreement among researchers who study the same field from its extreme far ends: the technical, and the philosophical. What we agreed upon as trivial is that the more computers forge ahead in their ability to replace human consciousness in various delineated contexts, the more pronounced becomes the fundamental theoretical difference between human and machine abilities, and this in turn highlights the dogmatic and rhetorical element in the writings of those populist prophets of technological innovation who herald the day when we will back up our brain and thus become immortal.

We talked some more, in detail, about the contribution of supercomputers to the field of translation and in particular about their contribution once their data include translations that are the work of humans and that are repeatedly corrected by end-users. The details of this conversation do not concern us here, but an anecdote I told them at the time can sum it up nicely. I told my fellows in conversation that I myself had already been corresponding reasonably successfully with a French colleague using the (still rather incredibly miserable) services of Google Translate, which are of course a very useful result (in certain limited and delineated contexts) of combining vast computational powers with the constant contribution of human beings who modify and improve the results on a daily basis (so that Google Translate is not, strictly speaking, a case of artificial intelligence). My colleague and I, I told them, succeed in corresponding rather well despite the amusing mistakes that this process generates. My letters to him, for example, invariably open with the words: "My very expensive friend…", a tradition launched by a mistake in the automatic translation of a simple sentence in French that means "My dear friend". The important point is that the computer's mistake was immediately embraced by both of us as part of a new speech tradition, and that it thereby ceased to be a mistake: it became part of the routine of our correspondence without anyone guiding us on how to do this and without any formal coordination. In this way, Google's automatic translator shaped our language and changed it through its mistake. To avoid misunderstanding I will emphasize that I have no doubt that the vast majority of translation problems of this sort will soon be solved, not because computers can talk like humans but simply because most of the things humans say to each other are pretty predictable and can therefore be matched with formulas in other languages that mean more or less the same thing. But the moral of the example is that these problems cannot *all* and *fully* be solved *as a matter of principle*. Language, as Bar-Hillel emphasized, is in motion, and no computer can fully predict this motion because its causes are drawn from all corners of our fractured *weltanschauung*. Obviously, then, so long as people are prepared to correct the computer's translation mistakes one by one and enter their corrections into the supercomputer, automatic translations will continue to improve, particularly in contexts that tolerate the occasional mistake and certainly whenever there are human interpreters at both ends of the translation who can easily correct any distortion (just as an intelligent reader might casually correct the mistakes made by human translators…). But it is no less obvious that there will always be some human beings who say truly novel things as well humans who understand new things from things that have been said before (every such act is a new translation of old texts in new and unexpected intelligent ways….). Therefore, automatic translation will never fully catch up with the sensitive human translator, nor fully predict the human translator's solutions to new problems of translation.

My new acquaintances and I continued sipping our coffee, which was very fine indeed, and watching the sunset, which descended suddenly and with unexpected tenderness upon the city rooftops as it sometimes does at the end of a long and eventful day. Even when a computer will be able to extract, output, compare and perform substitutions among the descriptions produced by all the poets who have ever witnessed a similar sunset, it will have captured nothing of the uniqueness of that enchanting moment.

Bibliography

Agassi, J. (1988). Analogies hard and soft. In D. H. Helman (Ed.), *Analogical reasoning, Synthese Library* (Vol. 197, pp. 401–419). Dordrecht: Springer.

Bar-Am, N. (2012). Extensionalism in context. *Philosophy of the Social Sciences, 42*(4), 543–560.

Bar-Hillel, Y. (1964). *Language and information: Selected essays on their theory and application*. Reading (MA): Addison-Wesley.

Bar-Hillel, Y. (1970). *Aspects of language: Essays in philosophy of language, linguistic philosophy, and methodology of linguistics*. Jerusalem: Magnes.

Bunge, M. (2003). *Emergence and convergence: Qualitative novelty and the unity of knowledge*. Toronto: University of Toronto Press.

Polanyi, M. (1951). *The logic of liberty*. Chicago: The University of Chicago Press.

Polanyi, M. (1952). The hypothesis of cybernetics. *The British Journal for the Philosophy of Science, 2*(8), 312–315.

Popper, K. (1950a). Indeterminism in quantum physics and in classical physics: Part I. *British Journal for the Philosophy of Science, 1*(2), 117–133.

Popper, K. (1950b). Indeterminism in quantum physics and in classical physics: Part II. *British Journal for the Philosophy of Science, 1*(3), 173–195.

Quine, W. V. O. (2008). Confessions of a confirmed extensionalist. Cambridge (MA): Harvard University Press.

Chapter 12
Revisiting Context and Meaning: Claude Shannon's Mathematical Theory of Communication

Sooner or later, every beginning student of communication comes upon some of the basic insights of the seminal theory that goes by the confounding name "The Mathematical Theory of Communication". As we will see below, Shannon's theory has distinctly formal attributes that are based on a typical extensional worldview of just the sort that we encountered earlier. The goal of the present chapter is to pose a question that straddles the line between didactics and methodology: How does Shannon's theory help students of communication understand their field? Or in other words: Should his theory be accorded a place in the simple introduction to communication, and if so, what is its role there?

The Mathematical Theory of Communication (which more properly deserves the modest title of The Theory of Information Quantification) was articulated by the brilliant electronic engineer Claude Shannon. It was given a more popular formulation with the help of the mathematician and administrator Warren Weaver. Weaver added to Shannon's theory reductionist pretensions that do not properly belong to it, and in particular, the claim to constitute the basis of a general theory of communication and even of the one and only complete theory of communication.[1] Thus, Shannon's theory came to be a fixture of introductory communication courses: still today, it is presented early on in these classes as a first and ultimately incorrect answer to the question of what constitutes communication and as a valuable but flawed preliminary blueprint for the proper way to study communication. In what

[1] A bit of history: working as an engineer at Bell Labs, Shannon completed the articulation of his theory during the years of World War II. He published his conclusions first in a classified Bell memo titled "A Mathematical Theory of Cryptography" (1945), and later in the Bell periodical publication under the modest title (and for that reason more characteristic of Shannon) "A Mathematical Theory of Communication" (*Bell System Technical Journal* 27(3), 1948, 379–423). The following year appeared a small book co-authored by Shannon and Weaver, pretentiously titled **The Mathematical Theory of Communication** (Urbana/Chicago: University of Illinois Press, 1949). The second part of the book is a re-print of the article from 1948, while its first part targets readers without mathematical training by offering a friendly and helpful summary of Shannon's theory as Weaver understood it. Yet in the first and third chapters of this influential introductory part of the book, Weaver expands the claims of Shannon's theory regarding the quantification and transmission of information to those of a general theory of communication, a theory concerned not only with the transmission of sequences of signals devoid almost entirely of content, but also with the reproduction of contents and intentions and even with their possible effect on different speakers.

N. Bar-Am, *In Search of a Simple Introduction to Communication*,
DOI 10.1007/978-3-319-25625-2_12

follows I will say some more about this particular way of presenting Shannon's theory so that we may later ask whether its status differs from the status of other theories taught in intros to communication and if so how.

Before turning to the basic tenets of Shannon's theory, let me emphasize that one main reason for our interest in it at this particular point in our argument is that its formal and extensional nature is especially pronounced. This is why it is widely perceived as a paradigm of the study of communication from a reductionist perspective. So, for example, according to Shannon, communication is successful when it results in the accurate reconstruction of a message—the latter perceived as a string of signals that a given source or sender seeks to transmit to a receiver—without regard to the question of how the message is interpreted by the receiver (or even if it has any meaning for the receiver or for anyone at all).[2] This conspicuously formal picture of communication is apparent in many other aspects of Shannon's theory that we will discuss below, such as his initially curious understanding of the concept of "noise" (transmission disturbance) as a message that carries the greatest possible amount of information. Shannon's theory therefore preserves and indeed underscores some of the central limitations that apply to any formal theory when it is regarded as a theory of meaning. And this is why Weaver's contention that Shannon's theory constitutes the basis of *the* theory of communication—the

(Footnote 1 continued)
Critics often note Shannon's consistently modest posture in the face of Weaver's overblown portrayal of his theory; but it must be said that in any joint publication, responsibility for the text rests fully with every one of its authors, and therefore that we cannot avoid the impression that, in his own introverted way, Shannon preferred to be equivocal about the elusive doubt surrounding the inflated reductionist claims attributed by Weaver to his theory. There is no doubt that in those parts of the book for which he was directly responsible, Shannon indeed emphasizes repeatedly that he is dealing with communication as an engineer, and that as such, he simply has no interest in meaning in general. But there is also no doubt that he avoided taking issue with Weaver when the latter argued that his theory could in principle be expanded to express, with equal success, all the semantic and functional-pragmatic aspects of the act of communication. On the matter of separating the formal aspect of communication (quantifying information), its semantic aspect (quantifying meaning), and its functional-pragmatic aspect (quantifying the possible influence of meaningful message upon speakers), Weaver writes: "[O]ne's final conclusion may be that the separation into the three levels is really artificial and undesirable" (Ibid, 25). See pages 24–28 for more on this point, and especially Weaver's summary on p. 28. This is, of course, the basic reductionist claim with which we are concerned here.

[2]The term "message" has multiple meanings and this multiplicity weighs heavily on many discussions of Shannon's theory because it stems precisely from the tendency, which we have here described as typical of the reductionist, to treat distinct levels of description as if they are virtually identical. Shannon's notion of a "message" can easily be misread as suggesting a meaningful message, i.e., as the content of a text, or even treated thus as the content of a message irrespective of its mode of expression, that is, as the speaker's intention alone, disregarding the way in which it is expressed. It is important to clarify, therefore, that Shannon's theory is concerned explicitly with the reconstruction of strings of signals and not of (meaningful) messages or intentions, and that Shannon makes this very clear. It is Weaver who (with Shannon's silent consent) attributed to this theory the claim to be the theory of the communication of contentful messages and even the theory of their possible effect on speakers.

complete theory of the accurate transmission of intentions, of narratives and their effects, and not just the theory of information quantification (1949, 24–25)—is so sensitive: it is one of the main incarnations of the reductionist claim that, in the final analysis, content is nothing but information (and this claim, in turn, as we saw in the previous chapter, implies the utterly confused claim that content is nothing but the potential for content. As we will soon see, Weaver's claim here is that content is nothing but a statistical relation concerning the relations between contents as it is perceived once we have forgotten the relation of the contents that it expresses.)

Since Shannon taught us both how to quantify information and how this ability helps us transmit it with optimal efficiency under changing transmission conditions, Weaver's claim implies that it is possible to do the same with contents, narratives, intentions and even their possible effects on various receivers in all possible contexts…. Now, this claim, as you already know from previous chapters, seems to me not only groundless but also expressive of a basic lack of understanding regarding the unique nature of communication. And it is here that the importance of Shannon's theory to the proper introduction to communication lies. Its role in such an introduction is dual, and crucial: first, it functions as a cautionary sign for all reductionists who seek to blur the line between a message voided of all content and a meaningful, contentful message, and second, it allows us to show how far we can nonetheless go in understanding communication from a reductionist point of view, while emphasizing the perpetual gap between them.

Now we are almost ready to describe Shannon's theory as it is typically presented in the literature (particularly of the kind that is historically accurate). But first it is worth pausing to acquaint ourselves with some of the theory's insights from a slightly different (more up-to-date) perspective. Consider a text. Of any conceivable kind. A text is a string of signals that constitute what Shannon called a "message"—prior to (or regardless of) any need or desire to transmit it. Shannon himself did not use the term "text" because he had no need to distinguish between a string of signals pure and simple, and a string of signals designated for transmission, but we will indeed distinguish between them—only slightly and only now—for the sake of emphasizing a central aspect of Shannon's theory. The texts we handle, of course, are made up of a finite number of pre-agreed signals. And it is helpful to notice that in every language there are agreements not just about the identity of these signals (agreements, for example, about the alphabet we use in English and about English-language punctuation) but also conventions, like habits regarding the frequency with which we use each one of them. These internalized habits characterize the particular language we are using. For example, in English, the letters E, T, A, and O are very commonly used: they are much more common than the letters Q or Z, for instance. Anyone who deals with deciphering codes knows that this difference is highly significant because given a good statistical analysis of signal frequency, it is possible to guess fairly well in what language a text is written, and this is a crucial step in its decoding. More interesting still is the fact that this is true not only with respect to languages in general but also, and often more emphatically, regarding different kinds of sub-languages, such as various jargons (like those of legal contracts, love letters, car manuals, etc.): they, too, are characterized by certain statistics of signal frequency. Moreover, it is not only the frequency of individual signals that characterizes a text and the language in which it is written, but also—crucially, for Shannon and all other code breakers—the frequency of particular sequences of two successive signals (known as "bigrams"). For example, the letter Q in English is almost always followed by the letter U, and after two successive vowels in English, the frequency of a third vowel is diminished (but not so in French). The fascinating fact lurking behind these dry considerations about signals and their "after effects" is that if we narrowed-down considerably the group of texts we wish to analyze—for instance, if we focused just on lease agreements in Manhattan—and if we carried out a highly accurate statistical survey of the signals featured in these texts and of their after effects, and if we then applied these considerations perfectly randomly to assemble a text (say, a text produced by a machine that knows no English), we would likely end up with a recognizable (if slightly amusing) replica of a Manhattan lease agreement. In other words, we would get a fragmented, garbled lease that reads like it was written by a very confused individual but which is nonetheless readable, at least in parts, and recognizable as a lease agreement. Knowing this amazing fact will be of use to readers of the present chapter, and in particular, will help their understanding of the incredible power of the theory we are about to explore.

Let us turn now to the traditional presentation of Shannon's theory of information quantification. Its basic insight is dazzlingly simple, and the world of the twenty-first century individual would be hard to imagine without it. Even if Weaver exaggerated its power, there is no doubt that the theory enabled Shannon and his followers—and we are all his students in many important ways—to make outstanding achievements: there is not a single action involving the cloud or your Smartphone, for example, that is not performed in light of and by virtue of this theory. Living in the rapidly-growing world of telephone communication networks, Shannon asked himself a practical question that can be articulated very simply: How can we improve the efficiency of transmission processes, processes that involve the transference of sequences of signals between a source and a receiver? As I said above, he did not draw a distinction between a text (a string of signals) and what he termed "a message" (a string of signals intended for transmission), for obvious reasons, and here I will generally follow him in this, except in places where it will be useful for our purposes to highlight various aspects that Shannon did not choose to note: thus, we will note that the message, more precisely, is simply a text that we wish to transmit to someone else. And another trivial yet crucial distinction: whenever we want to convey a message, we seek to do so through a given channel for transmitting information—a transmission channel. This could be an optical fiber through which our emails flow, for instance, or the transatlantic telegraph cable that once carried messages to remote countries, and even the evening air through which someone might try to see the light of a distant bonfire (announcing, say, a holiday, or the king's death). Shannon observed that the first step toward clarifying the challenge he tackled was to ask not what particular meaningful message (i.e., content) we seek to convey to others, nor even what particular message (in the formal sense of a set of signals) we wish to transmit, but rather what *sum total of messages* (what sum of texts) we *could* in principle transmit. This is because, all else being equal, the greater the number of messages we can potentially transmit in any given circumstance—that is, the larger the selection of texts (sequences of signals) that someone can in principle aspire to convey to someone else—the more complex becomes the task of actually conveying any given one of them (or if you will, the more complex becomes the task of successfully guessing the text, or accurately recreating it, at the other end of the communication channel). This observation is particularly useful because the ordinary channels for transmitting information are plagued by various disturbances (what engineers call "noise"), and overcoming these is a big part of the work of engineers. An example is the morning fog that might obscure the light of the bonfire mentioned earlier; or the random set of signals known as "white noise" that appear in various communication channels (most readers still know them as "static" on an un-tuned radio).

This seemingly simple basic observation allowed Shannon to do something that no one before him was able to do: to quantify information. Just like the grocer who weighs tomatoes or cucumbers, so Shannon could now "weigh" and compare various challenges of transferring signal strings under changing conditions (such as bandwidth, the set of possible disturbances, the level of accuracy required of the received message, etc.). This was a real revolution, because the quantification of information allows telecommunication companies to tailor optimal solutions to the various problems involved in the efficient transmission of information in different contexts, and to determine the costs of every such task in light of the various anticipated disturbances of communication—i.e., to determine the capacity of various communication channels. Quantifying information is simply attaching a numerical value to every message we might want to transmit, a value that accurately expresses its "weight", that is, the level of complexity of transmitting it (or, if you will, the theoretical complexity of reconstructing it at the other end of the channel). What Shannon emphasized is that the greater the number of potential texts we could in principle aspire to communicate—and hence of messages that the receiver rules out in the processes of identifying our message—the greater the amount of information that the message carries. This is because the number of potential messages that we now know that we did not receive rises—i.e., we had to rule out more texts in the process of determining what message we in fact were sent. And that's the whole of it, really.

The extreme cases that Shannon's theory describes are very helpful for understanding its basic principle as well as for understanding the limitations involved in viewing it as a full-blown theory of communication, that is, as a tool for the complete reduction of narratives and their meanings and effects in various contexts. At one extreme lies the redundant message. This kind of message is one that we have no need to communicate to others precisely because it is already known to the receiver before we even seek to send it. We may say, then, that the amount of information transferred between us and others when we transmit a fully redundant message, one that need not be transmitted because it is already known, is zero. At the same time it is worth observing (as Shannon was well aware) that the very existence of such a superfluous message presupposes the existence of communication conventions between the source and the receiver, and in particular, of agreements about what they both take for granted. Of course, Shannon does not discuss the question of how such a miraculous and indispensible foundation for communication can come into being out of nothing in the first place, how it emerges from the exchange of meaningless signals and nothing more. Thus, the redundant message in itself already points to the existence of a context that lies beyond Shannon's theory and which is not addressed within it.

The less probable the message, the more information it provides us with. Thus, at the opposite end of the spectrum from trivial information lies the least predictable string of signals. Note that the truly least predictable message is the one that lies outside the set of message that we foresaw in advance as possible: after all, we could receive a text that violates all the accepted rules for joining signals. That kind of message, Shannon's theory tells us, would contain the greatest possible amount of information. From a non-extensional point of view, one that acknowledges the existence of content expressed through the messages, that is, from a perspective that acknowledges the existence of contexts and meanings, we can say, further, that the message that exceeds the range of possible messages, the message with the greatest amount of information, sometimes (even if rarely) hints at the existence of a meta-message, and in any case often invites us to reconsider the terms of our communication. The implication of this kind of scenario is the discovery that what we had taken to be a comprehensive list of conventions for exchanging signals with others is in fact not comprehensive and needs to be re-evaluated. And it is important to see that a consistent extensional understanding of this scenario cannot be sustained, just like in the debate from the previous chapter concerning the Babylonian astronomical discovery, because the situation here is basically the same: in a consistent extensional system, the meta-message is not possible, and therefore neither is assessing the meaning of signs, to say nothing of an invitation to change the way in which we understand them. Thus, from an extensional point of view, the meta-message is received as a message like all other messages, but one that is not included in the list of possible messages, and therefore one that the system does not know how to process. (All-out noise cannot be communicated, of course, because if we wanted to communicate it, it would have to be a possible message in the list of possible signal strings and not a disturbance of their proper transmission.) Actual noise, then, is received as a fragment of a message that lies outside the range of possible strings that the receiver had anticipated. For Shannon the engineer, who wants no more than to weigh information against information and to transmit it reliably, such disruptions pose no real problem. He solved this extensional absurdity as engineers typically do: they simply pay no heed to the standard of consistency to which the reductionist will be held at the end of days. Thus, Shannon allowed himself to rely, modestly but nonetheless illegitimately from a strict reductionist point of view, on a separation of the context of his discussion into different levels, and this let him escape undesirable conclusions. He treated noise as an uninvited message that has blended into a desired message (and in doing so, he himself consciously blended together two different levels of description, the one describing the sum total of all possible messages, including noise as a possible message, and the other including only the sum of all desirable messages, excluding noise as a possible message). In this way, noise becomes both an anomaly and a predictable occurrence, that is, an anomaly that the engineer anticipates, understands and accounts for by retroactively removing its abnormality (just as all reductionist absurdities are explained away in retrospect) once the accordance between the original message and its received result has been checked and confirmed. As we have already witnessed, this is how the reductionist engineer solves all the problems that are unresolvable from a theoretical reductionist point of view (the point of view of the reductionist scientist). This solution is noteworthy, then, because it represents the crux of the reductionist response to the challenge posed by the existence of meaning, or in other words to the fact that any claim to a complete and comprehensive reductionist reading of a given situation is necessarily enabled by an act of self-deception (sometimes willful)—and more specifically, enabled by virtue of a context that continues to float about inside the engineer's mind, indispensable and yet unreduced. As I emphasized in earlier chapters, the ad hoc correction of a theory, its revision in light of discoveries, is always available and applicable as a methodological strategy, but as a comprehensive picture of the world it represents a genuine mistake, because any complete picture of the world, by virtue of being complete, must also explain the further context floating about inside the scientist's mind. For engineers, however, this is a desirable and useful strategy because they seek not a complete picture of the world but rather a solution to a given, limited problem. We must not underestimate the importance of these facts when we aim to assess the theoretical challenge involved in studying communication, because this theoretical challenge is thorough, and not ad hoc as in the case of engineering.

From Shannon's basic insight it follows that the simplest unit of information that can (and needs to) be transmitted is measured as a "bit" (a term that Shannon borrowed from his predecessors but which he helped popularize more than anyone else, so much so that he is often wrongly believed to have coined it). This point bears reiterating for the sake of highlighting a classic reductionist slip of the tongue: information does not contain bits and does not consist of bits, as many reductionists inaccurately put it, confusing different levels of discussion. Rather, information is measured, or weighed by means of bits: the bit is the measuring unit of the thing that is transmitted, not the thing itself, not the stuff that is transmitted and which we measure—just like a two-pound bag of sugar is not an answer to the question "What is sugar?" but just a simple way to compare the amount of sugar in the bag with other amounts. Similarly, "One bit" is never a valid answer to the question "What was just transmitted?" but only to the question "How much information was just transmitted?" (one signal was transmitted, of two possible signals). And what distinguishes the information that "weighs" one bit from trivial information, which weighs nothing at all, is that the source here selects a message to be transmitted from among two, and only two, available alternatives. In any case, that is how things appear from the point of view of the receiver at the other end of the transmission channel: the receiver cannot know in advance which of the two available alternatives the sender will in fact select until the message is received and deciphered as option A or B. And because from the point of view of the receiver the sender's choice is unknown until the message is deciphered, receiving the message (i.e., determining that it is such and not other) is also effectively a ruling-out of all the other possible messages, and this is why the information that the message carries can be weighed and quantified. What we weigh is the number of options excluded.

Here, again, we need to take note of the precariousness of the ground on which we stand from a purely reductionist point of view. For, even though Shannon never says this explicitly, his theory takes for granted that the existence of information (*even* the existence of information) presupposes the existence of some sort of environment—a kind of private point of view! Information exists because a system that is not omniscient, in other words, a system that in this respect is situated within a private environment, determines that certain possible actions (out of the sum total of actions that could be applied to it) did not occur. This fact is somewhat elusive but also very interesting philosophically insofar as it exposes the fact that Shannon's theory does not go all the way with strict reductionism. It exposes a fact that Shannon would not dream of denying but which a reductionist like Weaver, who seeks a comprehensive and consistent picture of the world, has no choice but to deny, namely that *the existence of information necessarily presupposes the prior existence of content*. This admission signals the reductionist's ultimate defeat: information, in Shannon's theory, is content whose meaning has been carefully minimized, and not the other way around: it is not the case that content is information "enriched" with context, as may well be the impression with which the previous chapter left some readers. (Realist reductionists, for instance, are convinced that since human beings are a thinking matter, the matter—that is, the absolute information, information that is totally free of meaning—must in the final analysis possess a spontaneous ability to develop an environment, a private point of view and a context, and it is indeed hard to disagree with them on this, since almost all of us are realists of this kind; but the question we are facing here is not whether realism is true but whether we will ever succeed in logically deriving all the properties of the private point of view from the properties of matter as such, and the answer to this question, we now realize, is negative.) These facts have vital implications for anyone who wants to study communication, because they mean that *it is in principle impossible to avoid tackling the notion of content as part of any comprehensive investigation of communication! We have to tackle it as a notion that is ultimately irreducible to that of information. We can run, and we can even run always and as a policy or principle (this is the reductionist strategy), but we can't really hide: we cannot give an exhaustive account of content sharing by means of information alone*. Yet this is exactly what reductionists ultimately aim to do: to expose phenomena of communication in the reduced world and to explain them all within a single, comprehensive reductionist account.

(Recall that in the introduction I said that I have a hard time with my colleagues' definition of communication as exchanges of information and of the means of communication as means for exchanging information. We can now appreciate why this kind of definition accentuates the problem we face and does not really solve it: information is an interim term, forever caught between the world of matter and the world of systems with an environment, but it is not a stable and complete bridge between these worlds. If the existence of information presupposes the existence of content, defining communication as a transference of information does nothing to help us remove the difficulty of treating contents from within our scientific picture of the world, because the most fundamental questions for researchers of communication are: *What is content? How is it shared? How can we study the exchange of contents, and in particular, how can we study content that precedes all of its retrospective descriptions as information?*)

This is why the ground on which we tread here is so precarious. We are now in a position to observe that even Shannon's theory, the undisputed and brilliant foundation of the reductionist treatment of content, the basis for the thrilling information revolution that the world is undergoing, is predicated, subtly but unavoidably, on a move that resembles sleight of hand or trickery more than an actual solution or even the promise of a future solution. Even this theory, we now see, assumes (in a sophisticated way) the existence of points of view from which occurrences in the world appear as possible choices, and later, in retrospect, as choices—and all of this is required not even for the sake of studying actual communication but just so that the study of quantifying information to improve the efficiency of its transmission can get off the ground! Recognizing this predicament has fascinating philosophical implications, and we have undoubtedly barely begun to understand them. The world of engineers in general and of communication engineers in particular is based on a very wise unspoken agreement to carry on as if the problem at hand belongs fully and squarely within an extensional realm as part of a consistent reductionist framework. In fact, however, every extensional realm is enabled in the first place by the prior existence of a realm of meanings, and this is just another way of saying that the rigorous reductionist framework is inconsistent —or, what amounts to the same: that the working framework of the engineer is not truly reductionist. Even in the case of machines that we regard as quintessential systems of information (like computers, for example) and which we purport to study as purely physical objects, we secretly regard them as systems with private environments, and thus as systems surrounded by content (the computer is a system that receives information and processes it, and in this respect we attribute to it an environment: it "responds" in similar ways to similar "stimuli", for examples to identical but distinct signals, and so forth). Only from a purely mathematical point of view is it possible to ignore the existence of environments completely, and this is precisely because the mathematician, more than anyone else, is aware of the question of what level of research and of description he is operating on at any given moment of his investigation. This point will be summarized one final time in the next chapter, so I will not go into detail here, but consider now, just for the sake of a brief illustration, the traditional reductionist, who was also a determinist. He aims to describe things, ultimately, from a divine, "retrospective" point of view, and from this point of view it seems clear that there are actually no real choices (as emphasized, for instance, by Spinoza). From this point of view—and this is the bottom line here—not only communication (the ability to convey and share contents) is an illusion but also the transmission of information à la Shannon! Thus, the consistent reductionist à la Newton must deny even the existence of information.

Shannon, as I mentioned, was not particularly troubled by these problems, and rightly so. He allowed himself to talk about information in terms of the "uncertainty" of a receiver regarding the "choice" of a source when discussing the communication between mechanisms that actually have nothing to do with mental states such as confidence, certainty, and choice.[3] And indeed, in the literature on Shannon's theory there is a very common tendency no to distinguish between the theoretical existence of a probabilistic option (an objective mathematical fact) and the uncertainty of a receiver regarding the way in which things have eventually turned out (a subjective psychological experience). I hope the philosophical background for this distinction is now clear enough. What Shannon calls "information", then, is from the very outset a hybrid of the reductionist and the emergentist pictures of the world—contrary to the impression that his reductionist adherents try to give us.

Shannon did not concern himself with the question of what meaning is, nor for that matter did he even really try to say what information is (although we saw above that he did make significant assumptions about the conditions for its existence). He was concerned only with the task of quantifying information in order to improve the efficiency of its transmission. In particular, he was not interested in the fact that information that "weighs" one bit can say very different things to different receivers and even to the same receiver in different contexts. But herein, of course, lies the perpetual gap between information and the context that restores to it levels of content that were artificially and retroactively removed from it. Reductionists tend to discount this gap by arguing that it can always be reduced. For example, they draw our attention to the fact that just as the number of signals (say, letters) is final and numbered in every language, so too the number of words in every language is more or less counted and sealed, and therefore anything that can be said can also be quantified while disregarding its meaning. But words, taken in and of themselves, are not contents, and here again the sleight of hand of the reductionist is exposed. The same word will express different contents for different speakers and in different contexts, and even for the same speaker in different contexts. The number of possible contexts, as we saw, is of course unlimited, and this is the crucial point here, for a word expresses a particular content only by virtue of a highly complex context, i.e., by virtue of its place within an individual's orientation activities (by virtue of that individual's purposes and the ways in which a given word is taken as furthering these purposes). The reductionist illusion is based on the idea that the sum of all possible texts somehow faithfully represents the sum of all possible intentions of speakers, and this is a deceit enabled by our willingness to overlook the fact that in order for a text to express a given content, let alone an original content, many more conditions need to

---

[3]Here is a typical quote from Shannon: "Can we find a measure of how much 'choice' is involved in the selection of the event, or how uncertain we are of the outcome" (49). This quote is intriguing because Shannon makes a point of placing the term "choice" within quotation marks, to emphasize that a receiver does not *really* have choices, but does not do the same for the term "uncertainty". Within the engineer's quasi-reductionist worldview (but not within that of the consistent reductionist), both choice and uncertainty can of course be expressed in terms of partial probability.

be fulfilled, some of which we can never guess in advance. The information that Shannon treats, then, is not content but content stripped of the vast majority of contexts and aspects that gave it its meaning, and it is clear that when a contentful message is left with only this thin aspect of its meaning, this meaning can then "round itself out" again in countless different ways depending on the countless contexts that we choose to spin around the message and depending on the system or entity for which it represents something. We can imagine, for the sake of an example, a kingdom with a dying king. At the top of the tallest tower in his castle stands a candle that will be lit if and when the king dies. To those who look up at the window of the tower, the un-lit candle sends the message that the king is alive. Lighting it sends the message that the king is dead. Now think of a different context: suppose that the king's daughter lights a candle in another window to signal to her lover that he may climb in through the window to meet with her in secrecy. The two cases present us with two possible message, each "weighing" one bit (candle A lit/candle A un-lit; candle B lit/candle B un-lit). From the perspective of Shannon's theory, the important point is just that the amount of information that passes between a source and a receiver increases in a simple logarithmic relation to the number of binary choices that a speaker can make when he joins binary signals together to construct a message; they thus expresses the maximum number of guesses that the receiver had to guess (if the receiver had to guess the information randomly, and so long as an accurate guess would be met with confirmation that it is in fact correct). Suppose, for example, that the princess' lover wants to learn both how the king is fairing and whether his beloved is available. He now wants to receive a message that weighs two bits, since there are four possible states in the world that the candles can represent as a system. A message that contains three bits of information represents a situation of eight such alternatives, i.e., information expressible with three candles, a message of four bits—sixteen options, and so on. Shannon's basic insight, then, is that when a receiver concludes that he has been sent one of sixteen possible messages, he receives more information than a receiver who concludes that he has been sent one of two possible messages. But a candle can in principle express countless different contents for different speakers. In particular, in our example, anyone who is not the princess' lover will have no inclination that the candle in her window carries any meaning at all.

The extreme cases we encountered here are described by Shannon as "redundancy" and "entropy": the redundancy of a message increases the more predictable the message, and its entropy increases the more it is unpredictable—that is, the more information it contains.[4] Shannon's work was of great value to the study of communication between computers because any given content is conveyable

[4]Noise, as we have seen, has maximum entropy in Shannon's theory. For some reason, this fact greatly impresses a growing number of postmodern writers, who tend, like the reductionist, to blur the line between a text and a message. To them, entropy is a metaphor for the subversive potential of deconstruction to tap new heights of meaning, or just for the aspired decadent fate of all cultures. Order (the antithesis of entropy), in this picture of the world, merely embodies the modernist's futile hope to give some structure and coherence to the world. For a brief survey of this surprisingly common trend, see Schweighauser (2014).

through various different messages, and these can be transmitted in ways that are more redundant or less redundant, and thus more and less efficient. This is where Shannon's theory goes from being a simple basic insight to a brilliant exercise in calculating the possibilities for compressing messages. For, Shannon's theory made it possible to calculate ways to maximize the efficiency of this process, and that is where its greatness lies. Consider, for example, the way in which spelling represents the phonemic expression of the spoken words. The phonemic expression of the sentence "I seek you", for instance, could also have been given the spelling "I Sik U" or "ICQ". And clearly, transmitting three signals is cheaper than transmitting eight. Quantifying the information, therefore, allows us to determine both the most cost-effective way to transmit information in a given timeframe and the most advantageous way of transmitting it securely in a given timeframe. Note that these are distinct problems, because noise, the big enemy of engineers that deal with the transmission of information, imposes certain limitations on the most profitable expression of information, especially when failure to transmit the information comes at a cost. The cost and the noise determine for us how much redundancy a message ought to contain because when an expression is economical to the hilt (as in the case of ICQ), in other words when it has the highest possible entropy, even the slightest degree of noise contamination disrupts the message to the point of rendering it unrecognizable at the receiving end, or worse, replaces it with a different possible message. This is, of course, a highly undesirable result when the substitution of messages is critical, can cause great losses, etc. Redundancy, therefore, is not a derogatory name for unnecessary signals: it is a fundamental factor in the stability of every system that transmits information. Without it, every random burp would make our friends' words unintelligible. Its proper dose, as you have probably already guessed, is necessarily context-dependent….

Recall that when we observe a message from Shannon's abstract point of view, and in particular when we observe it from the cryptographic point of view with which I opened this chapter, it is not just a string of discrete signals in which the probability of every signal is determined independently of all the others but a string of signals linked to one another through identifiable statistical relations representing the conventions of expression in a given language. If the signals were entirely random, if they did not in any way depend on one another, then every discrete signal would be a unit of information whose weight is fixed and a function only of the total number of signals in the language we are using. Take, for example, texts that are random strings consisting of the set of all Latin letters plus a fixed and pre-determined number of punctuation marks, like the space and the period. Unlike these texts, speakers of Latin or of English, as we noted above, do not join these letters randomly to form sentences: every language has clear structural and statistical properties, and those who have a mastery of these properties can typically assign every future signal a much more limited probability, especially when the context of the conversation narrows-down the number of words that can be chosen by the speaker and when the preceding signals are already known. General facts of this kind have been used for hundreds and thousands of years by codebreakers in their efforts to decode encrypted documents, with their only input being the language in which the documents are supposed to be decoded and maybe also, in some more or less vague sense, their broad topic (e.g., a report on the size and type of military forces, etc.). This is precisely why Shannon first called his theory "A mathematical theory of cryptography" and nothing more: the statistical wisdom he relies on is drawn entirely from the methods used by codebreakers. Thus, his information theory is based on the simple fact that all content-exchanges can be treated as if they are encrypted, i.e., as if they are exchanges of information. In that case, our only input regarding the content consists in statistics about the regularities that govern the relations between the various signals. But this input about the interdependence of signals is enough to enable the compression and secure transmission of a message, at a considerably reduced cost and effort. In some cases, we can even deduce how a particular message will continue even though we received only a small portion of it (in mathematics, these cases are described as "ergodic processes"). Shannon's theory, then, prompted enormous interest and progress in the study of these mutual dependencies and the processes of their prediction, and thus constituted not only the basis for the quantification of information but also, more importantly, the theoretical foundation for its optimal compression. The accuracy and the speed with which every computer responds to every command given anywhere in the world is calculated in light of this theory.

Most introductions to communication include some sort of description of the "Shannon Model of Communication". What they describe as Shannon's model is not a model so much as a general scheme made up of boxes linked by different arrows. The scheme simply expresses the fact that Shannon's theory perceives information as a linear flow of signals that originate with an individual entity (a person, for example), continues on to a transmitter (in a process known as encoding), then flows via a transmission channel to a receiver (in a process described as reception), and from there to its final destination (where it is decoded), while along the line that runs between transmission and reception there is room to posit one or another "noise" variable. Shannon posited this scheme not because he took it to be comprehensive and exhaustive of all possible processes of communication, but in order to emphasize that in the context of the very particular problem that concerns him—the attempt to quantify information and to compress it as far as possible for the sake of its efficient transmission—he is not interested in what human speakers understand from the messages they receive (in other words, he is not interested in meaning), but only in the ability of machines (a transmitter and a receiver) to transmit and receive sequences of signals at a desired level of accuracy. Yet as I noted at the top of the chapter, his friend and co-author Warren Weaver thought otherwise: according to Weaver, communication in all of its aspects conforms to the same basic structure that Shannon described, and therefore Shannon's theory offers a concise description of the core features of every possible act of communication, including the communication that takes place between two speakers without the mediation of machines. Even these speakers, according to Weaver, effectively function as transmitters and receivers when they turn thoughts into sounds and sounds into thoughts. And it is interesting to note that this view is still prevalent in the general public, including among many people who have never heard of Shannon and his theory, because it underlies the dubious PR rhetoric of the Artificial Intelligence community. We see the influence of this view in all the TED talks about robots who are said to have "intelligence" and "sensitivity" and in the futuristic predictions about computers with a moral sense replacing human judges and maybe also scientists and life partners. According to Weaver, what takes place inside the human minds of receivers when they understand the meaning of their interlocutors does not differ in any theoretically significant way from what the receiver-machine does when it processes a set of signals in light of statistical data that have been entered into it in advance. In both cases, Weaver believes, the receiver is asked to choose one item from a well-defined and limited set of possible items, and both cases involve data processing in light of clues of a deductive nature: the machine and the human are both at bottom systems that face a predetermined set of possibilities (whether these are purely formal messages or meaningful messages), and partial clues for choosing among them. Consider, for example, the pair of examples we discussed above involving information that weighs one bit: the case of candle A and candle B. Now recall their meanings in the contexts we described: the dying king, and the lover awaiting confirmation from the princess. Weaver argues that it is entirely possible that these two pieces of content are linked by some identifiable relation of dependence. It could be, for example, that the princess will be available to see her lover only if her father, the king, does not die, or that the king will die at once if he learns of his daughter's adventures. Therefore, just as identifying a particular signal reduces for us the likelihood of the appearance of another particular signal, so accurately identifying the relation between all possible contents allows us to improve the efficiency of their transmission. With this, Weaver re-articulates a staggeringly pretentious claim that we have already encountered—namely, that if we possessed the perfect divine theory, we could compress contents until they ceased to be contents because they would have been reduced to the mathematical function that describes their mutual dependence on one another, and nothing more. Had Weaver been right, Shannon's theory would indeed have reduced and removed the phenomenon of communication from the world. The content of what we say, in that case, would no longer have been mourning for the king's death or an eager signal from a princess to her lover but just the numeric representation of the statistical dependencies between these and all the other facts in the world. This is another incarnation of the same reductionist ideal that we already met; one more way of putting the claim that the qualitative and the quantitative are one and the same, that what we call "quality" is no more than a relation between quantities; and therefore, this is also the same familiar conflation of different levels of description that is so typical of the reductionist. It is the claim that the world of kings and princesses is an illusion, nothing but a world of candle flames that appear out of nowhere, disappear into nowhere, and signify nothing.

For readers who find my exposition so far to be clear and coherent, I recommend at this point skipping to the next chapter. The summary that follows is devoted to those who need more clarifications.

We encountered two very different senses of the term "information." In the first sense, we saw, information is synonymous with reality as it is in itself: it is the theoretical potential of this reality to become an environment (which is a sum of possible contents for a particular entity).[5] We called this information "absolute information", and we set it aside, because even though it is the foundation of every realist picture of the world and even though it helps us clarify the problem of consistency that plagues reductionists as a matter of principle, from a very abstract point of view, "information" in this sense is not discussed and indeed cannot be discussed in Shannon's theory: there is no way to quantify the full theoretical potential of reality to become content because the range of content possibilities embodied in every slice of reality is unknown as a matter of principle; we do not know and will never be able to know the sum total of all possible contents (this alone, of course, is enough to undermine Weaver's pretentious claim). The second sense we met of the term "information" is what we called "relative information." It denotes content that has been stripped of a significant part, but not all, of the aspects that render it meaningful. This information presupposes the prior existence of content, and is only achieved by means of a non-reduced *telic* context that embraces it. Therefore, any talk about fully reducing content to information represents a simple misunderstanding of the relation between these two terms.

In the chapter on extensionalism we encountered content stripped of many different aspects of its meaningfulness. There, the goal of this abstraction was to preserve and emphasize the deductive transparency of some of the distinct aspects of the original content relations: the extensional system allowed us to treat the content of expressions as if their meaning amounted to no more than a denotation of objects in the world (despite our awareness that this is not the case). The formal system goes even farther than this: it reduces these aspects as if they were no more than formal syntactical relations pre-defined within an artificial deductive system. The information expressed in these systems is therefore a measured and distinct presentation of (a limited sub-set of) pre-existing content relations. It is content that has been "slimmed-down" by the removal of nearly all the contexts that render it meaningful. The advantage of these actions lies precisely in the fact that this "diet" allows us to isolate particular aspects of the content and to get rid of others that are negligible from the perspective of a given purpose, as in some sort of sterile laboratory of meaning relations: this "lab experiment" helps us avoid deduction

[5]In the communication literature, the content that the environment provides the individual is typically given the ridiculously pretentious label "knowledge." The air conditioner, according to this view, has knowledge of the current temperature in the room and the amoeba has knowledge about the whereabouts of food in its environment. See, for example, Definitions 1, 10, and 17 in Schement (1993, pp. 20–22). Any first-year student of philosophy will tell you that we humans have very little knowledge, if at all, to say nothing of amoebas and air conditioners. Nonetheless, humans (like the amoeba and the air conditioner) are exposed to vast amounts of content. Content is conjectured knowledge and not knowledge in the overblown sense of the word as implied in the above definitions. In particular, to be able to convey information to one another we need not know anything. We need only to be able to share similar problems of orientation (or to embody them, as the air conditioner and the amoeba do) and to offer one another possible solutions to these problems while mutually agreeing, at least partially, about the procedures for choosing among these solutions.

mistakes of the kind that occurs when undifferentiated aspects of content become jumbled in our considerations and distort our calculations (in simpler terms, they help us avoid fallacies and computation errors). But we must not forget that this entire extensional discussion is predicated on the prior and unreduced existence of contents and contexts.

In Shannon's theory "information" is of course relative—that is to say, it designates very lean content and not a contentless material entity. Here too, then, we would possess no information at all if it were not preceded by content, a point of view, a subjective environment, and the like. This is why Weaver's claim that we can do away with the notion of content by means of the theory of information quantification is such a glaring error.

This point is very important because many other misunderstandings are inextricably linked to this mistaken view of Shannonian information as a genuinely content-free entity and even as liberating our scientific picture of the world from its built-in dependence on contents and contexts. This, as we have seen, is one more incarnation of the basic error of reductionists. It is important to note to our students, therefore, that when Shannon says that his theory is not concerned with meaning, we cannot take this at face value as implying that his theory does not depend on the prior existence of meanings or that it utterly ignores their existence. To see this we need only observe, as we did above, that there are certain non-trivial assumptions that Shannon takes on as a condition for treating information, assumptions about the nature of the thing that he is measuring, and without which he could not quantify this thing in the first place. First among these assumptions is the fact that the conventions of expression of the speaker and the receiver are shared and agreed upon by both. These conventions are of course a certain special kind of content that speakers coordinate among them before they can exchange information á la Shannon. Even the fact that every signal is necessarily part of a counted and pre-defined set of pre-defined signals underscores this prior agreement, i.e., the prior existence of conventions of communication that have been coordinated in advance. This alone, as we have seen, already distinguishes the information that Shannon refers to from absolute information, i.e., from reality as it is in itself, whose basic units we continue to this day to seek and will keep seeking so long as we are individual entities with private environments. Moreover, we saw also that the slices of information that Shannon treats maintain reciprocal ties of dependence, and that these ties can be quantified. Reductionists conveniently treat these as pre-given but meaningless relations that rise up out of nowhere like candle flames on the horizon, or more accurately, like the meaningless northern lights (because candles obviously do not light themselves). But unlike the northern lights, the statistical dependencies among signals express valuable human conventions: they reflect agreements between speaker and receiver regarding rules for the expression of contents that are of interest to them. The mutual dependence of the signals, then, is not accidental and does not appear ex nihilo with no guiding intention, even when it is presented retroactively as a dry quantitative analysis of the dictionary of a given language perceived as a purely physical object. In fact, a dictionary is not a physical object so long as it is treated as a dictionary, and *even a signal is not a physical object so long as it is a signal*! A signal is a platonic object (and therefore covertly but unavoidably intensional): it is the type or form that all token signals share, as perceived by those who view them as representations of the same signal, that is, by those who identify various tokens as different instantiations of the same form.

So, language speakers, individuals with private environments, have coordinated a dictionary of signs and communication conventions, and it is they who make it possible in the first place to quantify these signs and conventions in a way that largely (but not entirely) ignores the speakers' intentions. These conventions all express a unique type of intentions that we may describe as "meta-intentions", which in turn express a unique kind of content: "meta-content". This special content is concerned with the rules of communication, the rules for passing contents between speakers. When reductionists carry on as if these contents are nothing but senseless statistical relations between signals that have sprung up out of nowhere, they deceive themselves (and sometimes also their readers). In doing so, they ignore the fact that had these contents been mere statistical relations, the speakers could not have coordinated anything at all among them, and certainly not occasional amendments of these conventions according to context. But in fact we can do this: for instance, we can decide to amend our dictionaries. Such a decision is usually coordinated through the language that we speak. But even such a trivial coordination, as we saw, can never be captured properly and accurately by a one-dimensional extensional description of the situation. Of course, this coordination may well also affect the reciprocal relations of dependence that hold between the signals that represent these contents: language changes by means of our use of language. Meta-content, then, is content that relates to our shared conventions of communication, and without it, "communication" even in the sense that Weaver attributes to the word is entirely impossible. Awareness to the meta-content is what allows researchers of contexts to carry on as if they are studying context-free mathematical relations.

Shannon, as I said, was well aware of all of this, indeed more aware than anyone before him, so that the only puzzle in his case is how and why he agreed to add his name to a book that attaches Weaver's opening chapters to his own groundbreaking essay. Shannon never ceased to emphasize, for instance, that the context of a given discussion dictates the statistical relations between the signals, and therefore that any successful application of his theory depends first of all upon the accuracy with which the conventions that spell out this context are coordinated and defined. He always emphasized that these conventions, and even more so their amendment, is the most important piece of meta-information that speakers can exchange, the one that speakers have to take most care to convey accurately, because everything else depends on its accurate communication. But this supposed "piece of information" is not at all a piece of information: being meta-information, it is first of all a piece of content, and if it is transmitted as information, this could only have been done by introducing a meta-meta content. Of course, we could then go on to describe it as meta-meta information, but this move obviously never solves the underlying limitation once and for all.

This, then, is the root of both the greatness and the weakness of Shannon's theory as a general theory of communication: it succeeds in providing a faithful description (and even offers ways to improve the efficiency) of certain distinct aspects of the act of communication, while almost entirely ignoring the fact that it is dealing, ultimately, with an act of communication. Thus, it can only be successfully applied if its users are aware of its limitations and of the willful self-deception that it involves: it is their constant awareness that guarantees the theory's success. Shannon's theory shows us just how much we can achieve in the way of improving the efficiency of communication processes when we ignore (almost entirely) the fact that they are ultimately processes of coordinating environments among individuals. It allows us to consider information while seemingly disregarding its meaning (and in fact to consider it in abstraction from a significant part of the many aspects of its meaning, and herein lies the great achievement). And precisely because it illustrates how much we can achieve through this kind of reduction, the theory allows us, wherever it comes to our aid, to point out with clarity the very different nature of the activity it helps us measure. Thus, students of communication can and must make use of Shannon's theory, both to investigate their field and to understand the limitations of reductionism as a tool in this investigation. The theory helps them to identify those aspects of their field that need to be approached differently.

Bibliography

Arnold, D. P. (2014). *Traditions of systems theory*. London: Routledge.

Schement, J. R. (1993). Communication and information. In J. R. Schement & B. D. Ruben (Eds.), *Between communication and information* (Vol. 4, pp. 3–33). New Brunswick: Transaction Publishers.

Schweighauser, P. (2014). The persistence of information theory. In D. P. Arnold (Ed.), *Traditions of systems theory* (pp. 21–44). London: Routledge.

Shannon, C. E., & Weaver, W. (1949). *The mathematical theory of communication*. Chicago: University of Illinois Press.

Chapter 13
On Errors, on Correcting Them, and Thus on Goals (with and Without Scare Quotes): Cybernetics and Reductionism

It is time now to attempt to sum up the philosophical part of this book. We need to lay out its conclusions properly before turning to explore the possibility of applying them as part of a simple and friendly introduction to communication, or at least as part of the attempt to understand how realizable the dream of such an introduction truly is.

We have seen that an adequate introduction to the study of communication includes, above all and as a first step, a clear critique of the idea that a complete and final introduction to this field is possible, as well as an explanation of this interesting fact. This critique will help us outline the intellectual challenge posed by such an introduction: it signifies that the status of attempts to understand and explain communication is different than that of more isolated and "sterile" theoretical efforts, in which reductionism is the scientist's natural and only choice. *In the realms of communication, the observers and their environments are inseparable from that which they attempt to explain because every explanation potentially alters them and therefore may alter the object of explanation. Moreover, every change in the observers is potentially also a change in the communication dynamics that they create, and this changes their environments.* Similarly, new contexts give new meanings to old signs, thereby creating the need for new signs, and so on. This is why dictionaries always need to be updated. Communication, then, is a field that exists in constant flux, and the more the means for processing signs develop, the more our theories of communication become powerful, the more this motion is accelerated, over and over again. Pure and strict reductionism, therefore, is not a viable alternative here, as a matter of principle, and the claim that it is satisfactory fails to capture an aspect of the nature of communication that is so unique: the operation of self-orienting systems cannot be fully captured with reductive means alone, and the quiet assumption lurking in our picture of the world that reductionism is enough, or that we can and should ignore the fact that it is not enough, is misleading and obstructive. Like every scientist, the scholar of communication seeks explanations of his environment that are as strict and comprehensive as possible, but unlike other scientists, he must constantly develop, as an integral part of his training, an awareness to the limitations of these explanations and to the ways of handling these limitations.

N. Bar-Am, *In Search of a Simple Introduction to Communication*,
DOI 10.1007/978-3-319-25625-2_13

One way to understand how this awareness can be enhanced, which will also bring us into dialogue with the content of standard introductions to communication, is to explore the fascinating story of the field of cybernetics, founded by the brilliant mathematician Norbert Wiener. This story has an illuminating lesson for us, since cybernetics, like Shannon's theory, was presented by some of its enthusiastic adherents as a general theory of communication. Shannon and Wiener, it is worth noting, were colleagues who collaborated on several occasions, and their achievements indeed fit together to a large extent, as we will see. Cybernetics is a proposal to describe, explain and as far as possible study the possibility of controlling all orienting systems —from the air conditioner to the global market—as so-called feedback mechanisms (I will promptly explain what "feedback" means here and why it is so important to the study of communication). But like Shannon's theory, which exhibits different faces to the technician, whose interest is narrow, and to the scholar of meaning, whose interest is broad, so too cybernetics offers these two target audiences entirely different services. Wiener, too, like Shannon, offers invaluable practical knowledge to the narrowly-focused engineer, but as in the case of Shannon's theory, this knowledge is framed by an insight that embodies a highly valuable critique of the possibility that the knowledge will ever be expanded into a rigorous and comprehensive theory of communication. This represents a crucial critique of the very attempt to explain communication from within a purely reductionist framework. Thus, whereas the narrow technical foundation that Shannon and Wiener offer those scientists and technicians whose interests are narrow manifests only the various advantages of the reductionist perspective on communication, the general working framework they provide highlights above all the limitations of this point of view, that is, the theoretical limitations of the reductionist program as well as the need to complement it with other methods.

This double service is particularly clear in the case of cybernetics, which is why I have chosen to tell part of its story here. In its broad sense as the theory of all feedback systems, cybernetics was, for a fleeting moment in the latter half of the twentieth century, a dominant academic trend in the humanities and social sciences as well as in the life sciences and the (then budding) computer sciences. It even stirred the imagination of the military circles whose influence on the academic research agenda was then becoming dominant in the US and soon after in the whole world, as a result of which cybernetics enjoyed significant institutional support. The atmosphere that surrounded it in its early days was one of an intoxicating victory, which was sometimes wrongly presented as the final triumph in the struggle to explain communication. But when the broader picture became clear, it emerged that cybernetics actually underscores the theoretical limitations of uncompromising, systematic reductionism. And thus, rather abruptly, the charm of cybernetics lost its power: the field evaporated from the world of institutional research just as quickly as it had appeared. It was swept aside to the eccentric margins of this world, with the accepted wisdom being that apart from several of its basic technical insights, which are now part of the foundation of robotics and the study of artificial intelligence, cybernetics is no more than a collection of unsupported holistic aspirations.

I suggest that this fast zigzagging, between the announcement that cybernetics heralds the triumph of reductionism and its dismissal as holistic eccentricity, is an over-reaction of the academic world. Its origin lies in the fact that those who dismissed cybernetics did not properly distinguish between the empty holistic pretension to put forth a comprehensive theory of systems (a claim that some of Wiener's hasty followers did indeed attribute to his theory, despite repeated cautioning on his part) and the valuable critique that cybernetics offers of the possibility of a comprehensive reductionist picture of the world of communication (there is a silent mistaken assumption at work here—one that I come across quite often, and always with bewilderment—that every critic of reductionism is necessarily a confused obscurantist with a hidden occultist agenda, and there also lurks here an unjustified demand that every critic also immediately propose alternatives to the consensus challenged by his criticism). I think that the critique of reductionism that the cybernetic picture of the world entails, which I will detail below, has very important methodological implications for students of communication: it ought to dictate the proper structure of the simple introduction to communication that we are seeking. For *cybernetics does not invalidate the advantages of the reductionist approach to the phenomena of communication, but rather underscores the importance of this perspective alongside a firm recognition of its limitations, and thus the importance of the need to complement it with other means*.

So in what follows, I will recount briefly a piece of the story of cybernetics, both in order to give it its proper place in the simple introduction to communication and in order to draw from this story the relevant important methodological lesson. The methodological conclusions we will draw here concern the possibility of research that encourages constant criticism of reductive models in the social and life sciences, and they concern the ways to develop this kind of criticism. (We need to ask, in other words, whether and how responsible research that is not purely reductionist is possible in the life sciences and social sciences). After all, at least some of the theoretical limitations of reductionism have been known to researchers in these fields for some time now, but in the absence of an adequate alternative, the dominant tendency is always simply to ignore them as much as possible. What I would like to propose here, then, is a simple reform: *we should continue, as a matter of course, to advance reductive models of communication that are as strict and extensional as possible, simply because we do not really have any other models of explanation; but we ought, first, to advance them as models that are known in advance to be false (which is the exact opposite of how theories of communication are currently presented to students in standard introductions), and second, we ought to ask ourselves how best to develop the awareness of our students to the inbuilt shortcomings of these models. This is the question that I think ought to lie at the heart of the simple introduction to communication that we are seeking, and it is also the question that should lie at the heart of the teaching of communication (and of education more broadly).*

In the last part of this book I will try to demonstrate how this new critical function is recently showing up in many different sub-fields of communication studies, against the backdrop of fascinating new achievements in these sub-fields.

The excitement surrounding these discoveries is great, and rightly so. Since it is clear that biology, psychology, sociology, and even economics and history are all ultimately sub-fields of communication studies (or, if you will, of cybernetics…), and therefore that Wiener's lessons necessarily apply also to the researchers of these fields, who owe him the general framework in which they carry out their studies. After all, biologists, psychologists, sociologists and their colleagues from similar fields share with communication scholars all the same methodological difficulties that we have thus far described from a very abstract perspective, and in particular the inability (as a matter of principle) to bring any study in these disciplines to completion within a single strict and comprehensive reductive framework. The proper place of cybernetics in our general picture of the world, then, is heuristic: it allows us to recognize the limitations of the strict reductionist analysis in all the various disciplines that study communication and indeed hones our sensitivity to these limitations by pointing to the mutual, two-way dependency of different systems upon one another.

Before we unpack these claims, a note about Wiener's curious place in the academic world in which I myself was reared. In the philosophy department where I studied and in the department of communication studies where I later joined as faculty, his work was read by few, and he was almost invariably regarded as a thinker of very minor impact. When I began to study the annals of communication studies, I was bewildered by this fact, since it is hard to think about a topic more deserving of scholarly attention in the latter part of the twentieth century than communication, and hard to think of a philosopher of communication of a higher stature than Wiener. His absence from standard introductions to twentieth-century philosophy is therefore no less than scandalous in my opinion. For those who began their studies, like myself, in the nineteen nineties, it seemed like the man and his theory had been swallowed up by the course of history in a sinuous way: they became part of the general consciousness of academic celebrities, but not an actual part of the canon, of the standard "map of knowledge" that they teach. I suppose that one main reason for this is that when a scholar's contribution consists in presenting a subtle middle way between two contrasting worldviews, he is bound to strike both camps as confused, and thus as a confused representative of one's own camp, or worse, a confused representative of the rival camp. In his writings, Wiener indeed took great care not to come down one way or the other on any of the basic metaphysical questions that surrounded his theory, even when the inevitable implications of his position were clear to any sensitive reader. And so, cybernetics is often presented (even among some of its adherents) as an incredibly pretentious, almost outlandish attempt to establish a complete theory of systems, an attempt to reduce and solve within a single theoretical framework all the different problems of emergence that I have discussed here, or in other words, to establish a complete and comprehensive theory of communication. In previous chapters I explained why any such attempt is doomed to fail: the different problems of emergence are illustrations of a fundamental theoretical limitation of the reductive explanation, one that becomes all the more pronounced, and indeed cannot be overlooked, when we approach these problems as forming one grand unifying field, the study of communication. It is therefore vital to emphasize that Wiener was well aware of this limitation: for example, he emphasized the unavoidably goal-directed nature of his discipline, and thus the limitations that this imposes on our ability to reduce it fully to a more basic discipline that is totally free of such scientifically unappealing terms (Rosenblueth et al. 1943; Rosenblueth and Wiener 1950). He also emphasized that he did not seek to solve the different problems of emergence within a reductive framework—although he sometimes appears to be (or is presented as) trying to do just that. He repeatedly expressed his position that the goal of cybernetics is merely to provide the mathematical platform that will allow the study of systems qua feedback mechanisms. In other words, Wiener's goal was to emphasize the importance of investigating a variety of systems within a single mathematical and heuristic framework, a framework that allows us to assume their existence as systems without undue qualms, and thus, as a background to this description, and again without unnecessary inhibitions, to assume also the existence of goals…. (by contrast, the idea of a complete theory of systems, i.e., an accurate prediction of the behavior of systems in light of their environment and goals, presupposes also a detailed hierarchy of goals, and this is of course a truly paranoid idea). This error is especially conspicuous among followers of the biologist Ludwig von Bertalanffy, some of whom present Wiener as a disciple of Bertalanffy.[1] So I would therefore like to stress this one last time: if Wiener was right, if Polanyi, Popper, Bunge and other scholars with whom I am unequivocally siding here are right, we do not now and can not possess a strict and comprehensive theory of systems, that is, a general and comprehensive reduction of communication to a single theoretical platform, precisely because if we had such a theory we would have a complete and strict reduction of all goals to a single uber-purpose, and of all the different sciences, the different problems of orientation, to a single theory, and it makes no difference for the sake of this debate if we call this theory "the complete theory of everything", "a theory of emergence", "system theory" or just "basic physics". If these thinkers had it right, then supporting a picture of the world as consisting in multiple layers of systems means denying the possibility of a complete theory of systems, and "communication" is to a large extent the name that we have given to the impossibility of such a theory.

The fact that Wiener was a "middle man" whose nature and affiliation was not always clear to the participants in the heated debate between reductionists and emergentists has had a direct impact on the accessibility of his writings to readers and thus ultimately also on the penetration of his ideas into the academic canon and

[1]Bertalanffy's theory of systems has the merit of openly presupposing the existence of biological systems, and thus of multiple dimensions of existence facing the biologist who wishes to study a given ecological niche (Bertalanffy 1968). But in the final analysis, Bertalanffy proposes to reduce these different dimensions to a single level of description, thereby effectively cancelling them. Thus, *contrary to a common perception of his position, Bertalanffy's theory is opposed to the kind of general approach to systems that we are considering here, which denies the possibility of a comprehensive theory of systems*, that is, an approach that denies the possibility of a strict reduction of the behavior of all systems to a single level of description (whether it goes by the name "a general systems theory" or any other name).

among the general public.[2] Qualitative researchers were deterred by the mathematics in his work, while quantitative researchers overlooked it on principle (I am well aware of the paradoxical nature of the idea of "overlooking on principle", but there is no better way to describe the split attitude with which many reductionists approach cybernetics, if and when they approach it at all). This is true especially with respect to the fact—which Wiener took to be self-evident—that even though the study of feedback systems is carried out within a formal mathematical framework, it nonetheless assumes—indeed, takes for granted—the existence of goals or purposes, without which the mathematical description has no meaning: *purposes, then, are simply the informal contexts that make the formal discussion meaningful.* And indeed, Wiener's first presentation of cybernetics, in his first book,[3] was highly technical and unintelligible to many qualitative researchers, and in his second book (as well as in later appendices to the first book),[4] it was broad, accessible, and distinctly philosophical—and therefore impenetrable to many of his readers who were typical quantitative researchers and were alarmed by his general talk about a weltanschauung and about meaning. Of course, the difficulty of unifying these two presentations is the challenge with which Wiener grappled throughout his entire life: he sought to merge as far as possible, or at least to reduce the gap between the two realms that are unfortunately well represented by the two sides of the miserable rift between researchers of quality and researchers of quantity in the life sciences and social sciences. He sought out the subtle bridge between these realms because he understood that only those who look for this bridge are researchers of communication! (Only those who look for the bridge can recognize the gross mistake of those who argue that the other bank does not exist or that it is an integral part of their own bank; only they can recognize the tragicomic nature of ignoring its existence, as, for example, in the case of the reductionist who effectively nullifies his own existence as a scientist as well as a human being).

Wiener found that the best way to come closer to the building of this uncompletable bridge is by investigating feedback systems. He arrived at the study of these systems upon being enlisted in the American war effort in World War II. Wiener was asked to improve the algorithms governing the military's antiaircraft systems so that they would be able to calculate the precise location of a fast-flying plane when the missile intercepts its path, as opposed to the plane's location at the moment of calculation and firing (thus, Wiener is the great-grandfather of all

[2]This point is nicely explained by Ranulph Glanville in his panoramic essay on the evolution of cybernetics in Arnold (2014, pp. 45–77). Glanville also emphasizes the centrality of the notion of error to cybernetics (as I will do here), and in particular the fact that this notion is no less problematic for the reductionist than the notion of negative feedback, since the latter presupposes the former. While my main debt in developing this central point is owed to Michael Polanyi, and especially to his short and precise essay on cybernetics and his wonderful essay on the inability to reduce the study of life to chemistry, it is nonetheless pleasant to acknowledge also Glanville's helpful comments on this point (ibid, 52–53).

[3]Wiener 1948 [1961].

[4]Wiener 1954.

so-called "smart bombs" and "iron domes"). He succeeded in generalizing mathematical problems related to "strange" behaviors of these algorithms at critical points. And when he did this, he repeatedly explored, with the help of his expert colleagues who hailed form an incredibly wide range of disciplines, the value of investigating different phenomena (biological, psychological, even organizational and social) as embodying similar feedback mechanisms. And that is all of it really. From the mathematical point of view, of course, this is no small feat (in fact, a good part of Wiener's contribution to the scientific world lies in the mathematical details that underlie the general ideas presented here). But his great contribution to present-day students of communication has to do with the overall meaning of the picture of the world that he imparted to us. In the "modest", mathematical sense, there is no doubt that Wiener's theory conquered its targets; indeed, so sweeping was its victory that it is tempting to explain Wiener's relative anonymity as due in part to the outstanding success of his ideas: they became transparent by virtue of being suddenly taken as self-evident, by virtue of their incorporation into our world as the foundation of all the studies of robotics and of artificial intelligence. Nowadays, it is indeed hard to think about communication without thinking about the basic cybernetic analogy, which encourages us to analyze communication as a sequence of exchanges of information in the manner described by Shannon, supplemented by a feedback mechanism that allows the system to update its operations in light of changes in its environment, and by a meta-context that allows the researcher to understand these as functional corrections. Even our understanding of the air conditioner has improved as a result of this, since thanks to Wiener, air condition systems now feature not only an on/off button but also a thermostat that regulates the unit's operation toward a desired temperature. Thus, it was precisely due to Wiener's sweeping success that many of his reductionist opponents, as well as reductionists who mistakenly regarded him as one of them for reasons that we will soon discuss, tended to present cybernetics as the modern crowning glory of their picture of the world, as if cybernetics not only sits well with this picture, without real incompatibilities, but also guarantees the imminent reduction of the life sciences and social sciences to physics.[5] They did this by deceiving themselves:

[5]The two most prominent reductionists among the members of the small circle of early cyberneticists are the famous mathematician Jon von Neumann and William Ross Ashby. Von Neumann's work focused on the presumption to reduce human thought to the operation of computers, and Ashby's on the presumption to reduce the self-orienting organism to chemistry and physics. Wiener and von Neumann were well-known heroes of the early cyberneticists conferences, known as "the Macy conferences", but their relationship soured against the backdrop of the US military's growing intervention in shaping academic research after World War II, and especially in light of Wiener's growing disapproval of this tendency in general and in particular of von Neumann's increasingly close ties to the bodies and figures who propelled it. We may be justified in counting the influence of these elements on academic research among the main reasons that Wiener is less known today than he deserves to be: he himself predicted that he would no longer be able to publish any of his studies in light of his objection to this harmful trend, and even if he was wrong in this over-pessimistic prediction, he seems to have identified correctly the inclination to play down his place in the history of twentieth-century thought.

they embraced Wiener's extensional mathematical description while ignoring his most basic statement—namely, that feedbacks presuppose a *telic* context. Central to their move is the attempt to whitewash the concept of a "function" and legitimize it as a valid reductionist concept in the natural sciences by presenting it as a concept that is purely mathematical even when used by the engineer and researcher of systems. As I have already emphasized in various different ways in earlier chapters, no concept is harder for reductionists to handle than goal-directedness (or context), and none more crucial than this concept as a basis for the study of communication. For here we have the main bone of contention. Let us observe one final time why this is so by pondering over feedbacks.

Even a brief discussion of Wiener's theory must include an exposition of the concept of feedback in general and negative feedback in particular. But it is very useful first to pause and think about a more basic concept: the concept of "error" (or "failure"—I do not distinguish between the two here) in the most general sense imaginable. It is clear that the notion of error is central to the objects of many sciences. And what I want to stress here is that this concept is, of course, covertly *telic*, and for a rather trivial reason: there is no failure without a goal relative to which one (or something) fails (even the goal of failing is an intelligible goal in certain contexts, as is the goal of carrying on as if we have no goals whatsoever).

Given that the concept of error (or of failure) is unavoidably telic, *there is a basic difficulty in reducing all the sciences that handle errors, system failures, in all of their manifestations to sciences that we regard as more fundamental.* The difficulty exists even when we have no doubt (and we indeed have no such doubt) that what we see before our eyes is happening in full accordance with the laws of nature as described by a science that is in many ways indeed more fundamental. *The result is a layered picture of the world in which each layer consists in systems that obey laws that describe the more basic layer but also add to this reality elements that are on the one hand crucial to our understanding of our environment, of the behavior of complex systems, and on the other hand not reducible to the descriptions of the more basic layers.*

A few examples before we go on to generalize these points: consider, for instance, geneticists and bio-chemists who study errors in the replication of a given genotype. We know, of course, that from the "pure" perspective of chemistry, "replication errors" *qua errors* simply do not exist (in this sense, the reductionist is right to deny the existence of errors as a chemical event): this is because from the point of view of the chemist, the "successful" replication and the "failed" replication are both chemical events and nothing more, events that take place in light of the laws of chemistry alone and in accordance with them. What happens from this perspective happens for chemical reasons only, and a chain of causes and effects cannot in any way be wrong. Therefore, as Polanyi observed (Polanyi 1968), the view of DNA as simple information cannot be articulated in chemical terms, i.e., cannot be reduced to chemistry, since the chemist qua chemist does not even possess the necessary vocabulary for drawing a distinction between a successful and a failed replication of this information. Even Popper, who disagreed with Polanyi on almost every aspect of his philosophy, concurred with him entirely on this count: biochemistry is simply not fully reducible to chemistry, he argued, and this calls for a comprehensive methodological re-arrangement within biology.[6] For there is good reason for the biochemist's and geneticist's claim that certain events constitute a failed replication of the genetic material while others are successful replications: without such talk, we would have no understanding of the biological *function* of DNA and of mutations. Nor can the evolutionary biologist do without a basic concept of the "adaptedness" or "non-adaptedness" of animals to the ecological niche in which they evolved, and this is just another incarnation of the notion of "error" that concerns us here. Without a general concept of the adaption of individual entities to their ecological niche, after all, we have neither a Darwinian explanation of the origin of species, obviously, nor of any one of the organism's persistent and characteristic traits. As a matter of principle, therefore, the theory of natural selection and every explanation that presupposes it are irreducible to a more "basic" science. But of course, without the heuristics embedded in the principle of natural selection, we simply have no modern biology.

Anatomists who study the function of the kidneys or liver, and therefore also their malfunctions, as well as doctors who seek an understanding of the causes of a given disease and how to treat it, also clearly belong in this group. They presuppose the existence of a body capable of health, that is, of various manifestations of balance, as well as of deviations from these states: namely, disease and death. Consider also engineers, who discuss the success or failure of a smart bomb to hit its target or even the maximum load sustainable by a bridge, i.e., the moment when the bridge ceases to function as a bridge and goes back to being "no more than an aggregate of materials", as the physicist and chemist would describe it. Here, too, as in the case of the chemist and the biologist, it is plain that the physicist qua physicist and the chemist qua chemist simply lack the right vocabulary to tackle engineering problems—since the bomb that hits its target and the bomb that misses it both equally obey the laws of the world as described by the physicist; and the same thing obviously holds true in the case of the bridge. From the point of view of the physicist qua physicist, then, no bomb ever misses its target because bombs, as physical objects, have no targets to begin with: whatever obeys the laws of physics "misses" nothing at all. But this is not the situation of the engineer who wants to improve the performance of bombs—he *has* to think about them also as goal-directed systems if he is to make any progress in his research and design. Finally, consider sociologists and economists, who seek to understand how institutions function, why and under what conditions they emerge, and why and under what conditions they fail to perform their function (often against all the intentions and hopes of their founders). These scholars are clearly also studying systems and the breakdown of systems, or in other words—various institutions and their environments. And what is common to all the scientists we mentioned here is now becoming rather clear: they presuppose the existence (existence…! however loaded

[6]On this matter, see Popper's intriguing Medawar lecture 1968 included in Niemann (2014), as well as Niemann's helpful and interesting notes on it.

and problematic the meaning of this word may be) of the systems they investigate, such as the animal, the organ, the smart bomb or the market. And their aim is to study the conditions for the success of these systems. Cybernetics, then, was established so as to provide all of these scholars with the mathematical framework that would allow them to achieve this shared goal: to successfully study the goal-directedness of systems.

By now it should be easy to see why all the sciences we mentioned above are unavoidably *telic*—that is to say, they are purposeful not because they posit some sort of mysterious divinity that has bestowed on things an ultimate meaning as the propagandists for certain confused versions of the belief in intelligent design sometimes rush to conclude, but only insofar as they recognize that we cannot avoid describing ourselves as orienting systems with environments, that is without a context. Purposefulness here is the inevitable result of our self-description as orienting systems, and a precondition of our ability to identify an aggregate of materials as a system (with a function—that is, as a meaningful event in our environment). Aggregates, let me emphasize, do not make mistakes; they cannot fail in principle. Only systems fail, only they can have patterns of behavior and an environment.

And this is the basic point that Wiener saw as all-important: *when we regard a certain slice of reality as an orienting system, we are giving it a functional description whether or not we note this fact or appreciate its far reaching consequences*. And it is obviously better to note this fact, because from this moment on, the environment of that system is no longer, strictly speaking, the objective reality (and certainly not the objective reality and nothing more, as it is supposed to feature in the extreme reductionist's ideal description), but rather a certain (more or less) distinct aspect of this reality. *Indeed, the object of our observation is now a union of two distinct aspects of reality: on the one hand a physical reality, and on the other a telic space, that is, an environment, i.e., a translation of the physical description as the sum of all possible influences on the system's ability to approach its goal, to perform a certain function,* etc. These are once again the two faces of the same old coin and it is important to notice that it is now clearer why we can neither ignore either one of these faces nor, of course, eliminate either one of them permanently.

The possibility of studying system failures, then, necessarily involves acknowledging the existence of goals, just as acknowledging the existence of systems requires us to acknowledge the existence of their environments. Moreover, these two pairs are inextricably linked, because purpose or goal is that which creates an environment: by virtue of the goal, certain distinct aspects of reality are unified functionally as similar stimuli that therefore invite a similar response from the system, a response pattern. *Without goals, then, there are no systems.* And so, anyone who posits the existence of some sort of system thereby posits the existence of its environment, and also relies, whether consciously or not, on a *telic* explanatory framework. It follows from this that some goal-directed context will remain an integral part of the scientific picture of the world so long as we wish to describe our world as populated by systems.

It is important to note that much the same applies also to the concept of "problem" (or "challenge" or "question") in its most general sense, as it is defined for us in the logic of questions. A question, in this logic, is defined as a set of mutually exclusive alternatives along with an invitation to choose one of them as true. I note this here because this fact has critical importance in developing the positions that I am outlining here as part of a comprehensive philosophy of communication. According to the theory of questions, "question", "problem", and "challenge" are more or less synonymous terms because they all denote a set of alternatives (action courses are but an instance) laid out before the system along with the invitation (or demand, or need) to choose one as correct. And as we know already, environments are spaces of competing courses of action and their consequences. Therefore, the existence of systems is inseparable from the existence of problems of orientation. Moreover, the discussion of problems of orientation is the foundation of any discussion of systems and their environments and therefore of any attempt to study communication in a general way. Even objects to which we do not noticeably attribute an active ability for self-orientation in an environment, like a bridge for example, are systems in this modest sense by virtue of the ability of our consciousness to attribute to them a function, that is, to see them as part of a problem or challenge, as an attempt to achieve a certain goal. Thus, in the case of the bridge, we carve up the countless possible aspects of (objective) reality into a set of events under which the structure will collapse and cease to function as a bridge (in other words, the aggregate of materials that we identify as a bridge will no longer do the job expected of it as a bridge) and another set of events under which the structure will continue to serve its purpose (namely, not collapsing under such-and-such weight).

You may be tempted at this point to ask one final time: if the "soft" goal-directedness of all the sciences dealing with systems is indeed so basic and clear-cut that none of them can take even a single step without presupposing environments and goals, and thus the inconsistency of the reductionist picture of the world, then how can the reductionists deny all of this with such assurance? Well, it seems to me that reductionists have only one final and rather flimsy defense here, which we should note before we move on: they describe these sciences as "applied" sciences and therefore as sciences that are not "pure". In particular, reductionists regard the ontology of the sciences mentioned above (as indeed they must if they seek consistency) as "a manner of speaking", a convenient, loose style of expression and no more. Thus, they contend, from these sciences we cannot conclude anything about (the essence of) reality as it is in itself. The model here for the reductionist is clearly that of engineering: the engineer qua engineer, claims the reductionist, allows himself to talk and carry on as if smart bombs and their targets were actual things in the world, but in truth (qua physicist), he knows that they do not exist, that there are neither bombs in the world nor targets: there are only convenient abstractions that we view, momentarily, as designating distinct ontological entities, for the sake of convenience and in particular as a means of applying physical knowledge to engineering problems. A more elaborate and detailed answer than the one we offered here to the claims of reductionism exceeds the aims of this book, since it involves a meticulous discussion of the role of an ontology in the scientific endeavor as well as of the question of what a description of the world that includes no systems of any kind would look like—or in other words, what a complete and consistent reductionist explanation of reality could look like. These are issues whose treatment even in the most current literature that I am aware of is very disappointing, a state of affairs that illustrates how harmful the uncritical acceptance of reductionism truly is to genuine debate about the foundations of scientific explanation. But by way of summing up, I will note one last time that the reductionist may be invited to point to some field of physics that does not require these convenient abstractions, and also that he should apply the same conclusions that he draws in the domain of engineering (as described above) to biology and to psychology—that is to say, he should similarly deny the existence of animals, including his own existence as an individual animal, as well as his existence as a thinking being, the existence of his thoughts as a thinking being and the existence of any content to those thoughts: all of these, he is committed to claiming, are strictly speaking no more than illusions. This denial, in turn, leads inevitably to a denial of all the scientific problems that concern the non-existent scientist and which he set out in the first place to resolve. This is not a simple predicament to say the least, and seems indeed to be paradoxical, because the only way an individual with purposes can, as part of his purpose, deny his existence as an individual and the existence of his purposes (including the purpose of presenting them as non-purposes) is as a *façon de parler* (and a rather sloppy one at that)…. And this, of course, is just the opposite of what the reductionist intended.

Now that we have adequately legitimized the unapologetic use of modestly purposeful language—not the kind that assumes the universe was created so as to complete some mysterious divine goal but, rather, language that reminds us that all scientific investigation is embedded within an unreduced context—we can finally turn to consider the most basic concept of cybernetics—feedback—and the contribution of the picture of the world that it facilitates to the study of communication. For Wiener's proposal was, at bottom, to study all orienting systems as far as possible through the mediation of this concept—that is, to aim to reduce orientation and communication to the constant operation of feedback systems. Wiener emphasized that *feedback is not really a concept that the strict reductionist can be happy with, as it is semi- or quasi-reductionist: it does not allow us ultimately to get rid of systems and their goals in favor of a picture of the world that avoids them. For the description of feedback as such* (unlike its description as a purely physical event or its purely mathematical description in a formal framework) *necessarily involves a two-tiered description* (at least) *of the given events: as physical and (at the same time) as telic, that is, as physical events on the one hand, but at the same time as stages in the movement of a system toward or away from its goal, on the other hand.* The concept of feedback, then, is the product of a conscious and deliberate combination of two distinct levels of description of events—their description as free of meaning, within certain known limits, and as meaningful, and fully grounded in a given context. *Without this blending of levels, without this kind of two-tiered descriptive structure, we cannot study feedbacks, to say nothing of communication, as a series of feedbacks (and perhaps cannot study reality at all, as Polanyi argued).* Now, allowing this sort of two-tiered speech of course puts an end to the zealous reductionist dream of a final theory that describes the various different phenomena of our world (and communication in particular) using a single level of description. It is only through the multiplicity of descriptive layers that we are able to breathe life and meaning even into extensional descriptions of reality and to view them as expressing the more or less successful conduct of systems in their environment, conduct whose amelioration is the object of communication studies.

This fact is particularly conspicuous in the case of negative feedback. For *negative feedback is the correction of a system's behavior in light of a deviation that occurred in its movement toward its goal. Its conduct, then, is from the very start perceived in a two-tiered way: both as part of a chain of physical events—the physical effect of a physical cause occurring in a physical world in which there are no errors and which lends itself to an extensional or formal mathematical description—and as part of a chain of telic events, a correction performed by the system in light and by virtue of its deviation from its goal.* This is the living, breathing heart of cybernetics. Reductionists seek to play down this breathing heart with the argument (undeniably true) that the heart—or by analogy, that feedback—is nothing but a bundle of particles subject to the laws of physics. But in order to describe the heart as a heart we have to describe it also as a vital pump within the body, that is, as having a function, and in order to describe the feedback as feedback, our view of reality must be (at least) two-tiered, and in particular, we have to view reality as embodying different stages of the goal-oriented behavior of system in their environment. Reductionists want to focus exclusively on the first description, an extensional or formal description of the events as lacking any further meaning. But were they truly able to remove the further meaning of things, to get rid of the intimate link between goal-directedness and our view of the world as made up of systems and their environments, this glorious reduction would thereby also lose its value. The feedback qua feedback would also be lost, and with it Wiener's entire wonderful quasi-reductionist achievement.

Positive feedback is less complicated in this respect, but precisely for this reason it is hard at first to observe that it, too, is silently *telic* if only in the modest sense of a propensity or a tendency to achieve or maintain a certain state. Consider a very simple case, one that does not even involve any real case of orientation. Imagine a snowball rolling down a hill. It has no real goal, obviously. (On this point Spinoza famously made the witty reductionist remark that if a rolling stone had consciousness, it would believe itself to be rolling freely, of its own volition… a metaphor that probably rolled into Bob Dylan's famous song and from there on to the famous rock band). But we can also view the snowball's gradual increase in size, if and when it occurs, as a propensity or a tendency—i.e., an inclination to proceed in a certain direction. We would do this simply to allow ourselves to pose the question of whether or not the snowball's increase in size will persist, whether it will cause another chain reaction down the line, leading in turn to a further increase in the snowball's size, and so on. Thus, even a purely extensional or formal description of the snowball—for example, as a series of numbers representing sizes at given moments—is suddenly perceived as also the description of a propensity, and thereby also as embodying an answer to a scientific question that assumes that reality expresses various kinds of such propensities, i.e., a question that assumes a very modest form of directionality, which is no more than movement toward and away from an abstract target. *And so, when one event in a chain of events leads to another event, which is perceived also as progress in light of a certain end, we have a case of positive feedback* (the snowball's increase in size, repeated over and over again). Every bomb and every landslide—a landslide of rocks, a financial landslide, or even the collapse of body systems—is thus a case of positive feedback. An even more familiar example, which already clearly includes a system for receiving and processing transmissions, occurs when an electronic amplifier receives audio signals that it has itself put out, and thus amplifies them, over and over again. This, then, is the positive feedback: a pair of events perceived both as a cause and effect from the point of view of physics (or other sciences considered as basic), and as the constant and increasing movement of a system toward its goal from a goal-directed point of view.

Negative feedback, by contrast, occurs when one event in a chain of events (the cause) is perceived as a deviation from the chain's trend, and another event (the effect), is perceived as a correction of this deviation. It is a pair of events perceived both as a cause and an effect, and as a deviation from a desired course and its correction. And that is all. Indeed, because there is so little to this idea, it can be hard at first to grasp the breathtaking scope of the metaphysical revolution that it smuggles, ever so modestly, into our scientific picture of the world: *the need to describe the world using the terminology of feedback is a quiet admission of the ultimate limitations of reductionism as a comprehensive picture of the world.*

Negative feedback of course foregrounds our concerns as researchers of communication because the correction of errors, which underlies negative feedback, is distinctly goal-directed. This makes it easier to realize that we are no longer in the world of the strict reductionist when we apply this concept. Wiener's great metaphysical revolution, then, hides itself as the simple looking claim that feedbacks are indispensable for investigating orientation and communication, and thus that no strict reductionist description of these phenomena would ever be complete. Negative feedback (and in particular its combination with positive feedback, as in all advanced cybernetic systems) allows the discrete bits of matter of the physicist to partake in goal-directed behavior, and thus to become meaningful. This could appear like a miracle of emergence, and maybe also, mistakenly, like a solution to the problem of emergence (because thanks to Wiener, the operation of the mechanism of negative feedback is described in ideal cases with satisfactory precision by the engineer…)—had we not already noted that feedback is itself by no means a strictly reductionist concept. So that when we say that the air conditioner maintains a more or less steady temperature in the room in light of changes in room temperature, i.e., that the air conditioner responds to temperature fluctuations in a way that keeps the temperature in the room more or less constant, we are seeing it as a cybernetic system (a servomechanism), and this already uses language that no strict reductionist who aspires to consistency can accept. This, by the way, is also why we call the smart bomb "smart": it can re-direct itself toward its target even as the latter is trying to get away from it. Positive feedback mechanisms, or more accurately a delicate interaction between these mechanisms and negative feedback mechanisms, enable the smart bomb to accelerate its movement toward its target as well as to explode successfully, while negative feedback mechanisms, or their delicate interaction with positive feedback mechanisms, allow it to re-direct itself toward the target even when the target is in motion, and indeed in evasive motion.

Cybernetics was a cause of great excitement among reductionists, and with good reason. We are now in a position to understand this excitement, for it is quite obvious that when a host of phenomena that had appeared to be mysterious and which concern orientation and communication, life and consciousness, is suddenly described and studied through mechanisms that we are able to describe from within the perspective of engineering, and even from a pure mathematical perspective, then it would seem as if the problem of reducing these phenomena has been significantly minimized. This, at least, is how things seem until we notice that a real gulf separates the quasi-reductionist world of cybernetics and the one-dimensional, feedback-free world of the strict reductionist.

Bibliography

Arnold, D. P. (Ed.), (2014). *Traditions of systems theory*. London: Routledge.

Bertalanffy, L. (1968). *General system theory*. New York: Braziller.

Buckley. W. (Ed.), (1968 [2009]). *Modern systems research for the behavioral scientist*. Chicago: Aldine Publication Company.

Conway, F., & Siegelman, J. (2005). *Dark hero of the information age: In search of Norbert Wiener, the father of Cybernetics*. New York: Basic Books.

Niemann, H.-J. (2014). *Karl Popper and the two new secrets of life*. Tübingen: Mohr Siebeck.

Polanyi, M. (1952). The hypothesis of cybernetics. *The British Journal for the Philosophy Of Science, 2*(8), 312–315.

Polanyi, M. (1968). Life's irreducible structure. *Science, 160*(3834), 1308–1312.

Rosenblueth, A., & Wiener, N. (1950). Purposeful and non-purposeful behavior. *Philosophy of Science, 17*, 318–326.

Rosenblueth, A., Wiener, N., & Bigelow, J. (1943). Behavior, purpose and teleology. *Philosophy of Science, 10*, 18–24.

Wiener, N. (1954). *The human use of human beings: Cybernetics and society*. Boston: Houghton Mifflin.

Wiener, N. (1956). *I am a mathematician: The later life of a prodigy*. Cambridge, MA: MIT Press.

Wiener, N. (1948 [2nd expanded edition: 1961]). *Cybernetics: Or control and communication in the animal and machine*. Cambridge, MA: MIT Press.

Chapter 14
Back to the Basic Problem of Communication: The Limitations of Cybernetics

Cybernetics is usually mentioned in the same breath with the pioneering work of Cannon (1932), and indeed as an elaboration of this classic work.[1] Wiener, in his autobiography, notably describes Cannon as the leading American scientist in the interwar period, and it appears that an early encounter between them when Wiener was still a boy left an indelible mark on him.[2] As early as the nineteen twenties, Cannon had already begun to realize his life's work: describing the human body as a complex of balancing mechanisms that maintain relative constancy (homeostasis). These mechanisms, he observed, achieve and maintain various aspects of stability through moderating responses to deviations from this stability.[3] This moderating response is, of course, a crude precursor of Wiener's negative feedback. Thus,

[1]Cannon's book *The wisdom of the body* was published in 1932, but his studies on homeostasis (and his first uses of the term) began to appear already in the nineteen twenties. His collaborations with Wiener took place largely through the mediation of the brilliant physiologist Arturo Rosenblueth, who was Cannon's foremost student and his co-author in some of the early studies on different mechanisms of internal regulation in the human body and in the nervous system in particular. Only later did Rosenblueth become a close friend of Wiener's and a key figure in cybernetics. Wiener and Rosenblueth co-authored various important articles in a variety of fields, from the methodology of science to the physiognomy of the nervous system. Most prominent among these for communication scholars are Rosenblueth et al. (1943), and Rosenblueth and Wiener (1945, 1950).

[2]An unduly short account of this meeting, intriguing in its baffling brevity, is provided by Wiener in his autobiography (Wiener 1956, p. 171). On Cannon's place among the scientists of his time, see p. 221. From reading the various details about Cannon that Wiener scatters throughout the story of his life it seems almost as if he revered Cannon to the point of fearing full collaboration with him, a fear that dissipated only when it was too late for any real collaboration given Cannon's advanced age. The two men met to discuss their work only close to Cannon's death and without resulting in any significant publications.

[3]Precedence arguments about the first source or instance or appearance of ideas whose evolution is clearly gradual are usually pointless if not ridiculous, expressing more than anything else the narcissistic hope to be born ex nihilo, out of the white foam of the waves. Cannon and Wiener are inspiring examples in this respect, since both were at once impressively original and very generous —indeed, excessively so—in attributing this originality to their predecessors. The first chapter of Cannon's book is devoted entirely to quotes by his various harbingers, most notably the great Claude Bernard. Cannon notes many other sources of inspiration but also adds wisely that the basic idea of human health as the ability to maintain a relative balance between competing forces can be attributed even to Hippocrates. Beyond his generosity and humility, Cannon was keenly

N. Bar-Am, *In Search of a Simple Introduction to Communication*,
DOI 10.1007/978-3-319-25625-2_14

Wiener gave clear mathematical meaning and a technical vocabulary to the insights that Cannon had expressed in a rudimentary and intuitive way. But Cannon's great credit is that he emphasized the fact that what we see and describe as our body's sensitivity to its environment, in other words, as the body's orientation in its environment, is expressed simply as the maintenance of various aspects of equilibrium or homeostasis in the face of intrusions.

For example, our body maintains a more or less constant temperature partly through rather simple feedback systems: we sweat when our body temperature rises above the desired level, and we get goose bumps when it drops. In this way (and in various other ways, of course, like shivers, changing the body's position, etc.), the body maintains a more or less constant temperature when its environment changes. In this, our body is, as it were, halfway between activity and passivity, between operating in light of its environment and being operated by it. Our skin cells, for example, expand and shrink as a result of their warming or cooling, so that the sweating and the goose bumps are both the body's response to the change in temperature and its passive activation by this change. What we have here is one more way to underscore the two-tiered nature of every case of orientation that we see before us. Orienting oneself in the environment and being activated by it are the two sides of the same coin we encountered earlier: it is the basic wonder that researchers of communication seek to understand. Thus, Cannon's search for a theory of homeostasis is a highly important precursor of the search for a simple introduction to communication. (The body, by the way, is able for a short time to maintain internal thermal equilibrium even when its external environment is rather extreme—for example, 200 or −30 °F—and this fact is especially notable given that we are made of water, protein, and the like, so that the body is here clearly performing a task that its separate components cannot handle: the system indeed transcends its parts, in this sense.)

Maintaining the relative temperature of our body is just one example among a vast number of examples that Cannon assembled and studied throughout his life. Our body, he showed, is an incredibly complicated system of homeostatic mechanisms linked to one another through breathtakingly complex hierarchies. It is these mechanisms, as he demonstrated, that enable and sustain life: through them, the concentrations of fluids, minerals and hormones in our body are balanced, thanks to them the nerves feel and the eye responds to the amount of light around us, through them our muscles keep moving without curdling like egg white, our wounds heal, and our immune system wards off various threats. The words "through" and "thanks to", however, are imprecise in this context, because the body, as we just noted, is both identical with these feedback systems from a material perspective and exceeds them or "uses them" from a cybernetic point of view. This is, once again, that

(Footnote 3 continued)
aware of the clash between the basic idea underlying the theory of homeostasis, on the one hand, and reductionism on the other, so that his list of distinguished forerunners may well be an attempt to cushion the impact of this clash on his fiercely reductionist milieu.

two-tiered feature of our descriptions of systems that we revisit from different angels, and it is a feature that first became clear and distinct thanks to Cannon's work.

However, Cannon's theory is limited in comparison to Wiener's, and not only from the formal mathematical standpoint, since Cannon focuses on the body's tendency toward relative stability, he gives no real place in his theory for "positive feedback". Obviously, not everything that happens in the body is best described as a stabilizing attempt, even if we expand this definition to include failed stabilizing attempts. Consider, for example, the body that goes into a state of shock from blood loss. This is clearly a failed homeostatic response as part of the body's attempt to prevent too much bleeding: the blood pressure drops to avoid an excessive loss of blood. But it can become part of a positive feedback loop, with the drop in blood pressure in turn causing inadequate blood supply to various body systems, thereby sending the body into a state of "shock". The "shock" can then trigger further chain responses and thus elicit multiple organ failure and death. What we have in this case is first of all an occurrence that deserves to be described as a positive—rather than negative—feedback, since a failed stabilizing attempt is still not a deteriorating dynamic, and our main focus here is on a process that leads from one deterioration to the next. The absence of positive feedback from Cannon's theory is exceptionally curious considering the fact that he arrived at his studies of homeostasis through his famous study of the physiology of shock and of flight and fright states, a subject that preoccupied him throughout his career. As we will see below, the lack of a distinct and complementary concept—positive feedback—is expressed in various areas as a heuristic lacuna in his system: the concept's absence suppresses a whole set of guiding questions that are vital for researchers who seek to use the system-oriented picture of the world, which he helped create, in order to understand various phenomena.

In any case, Cannon is a central figure in our story not just because he is the venerated grandfather of the idea of negative feedback. He is central to our story first of all because he already clearly articulated the possibility of *a general theory of homeostasis*, making him a lucid and highly influential harbinger of the idea of a general theory of communication, if only as an ideal aspiration for the future or, better, as a heuristic proposal for the scientific agenda of those who wish to follow in his footsteps. For *he called the attention of his colleagues to the fact that homeostatic mechanisms are all cases of a single cross-disciplinary phenomenon, visible from all corners of the life and social sciences, and that studying all of these mechanisms as cases of the same phenomenon stands to benefit each separate discipline and should not threaten or alarm scholars in the exact sciences.* Cannon's predecessors, in fields ranging from psychology and sociology to ecology and economics, had already identified homeostatic mechanisms under many different names in their separate areas of specialty, but his important contribution was to emphasize the heuristic benefit of noting and studying their trans-disciplinary similarity. This is why he is the herald of the search for a simple introduction to communication.

Before we move on, it is worth mentioning a few examples to underscore the scope of the interdisciplinary metaphor that concerns us here, and its utility. Even

Freud, for example, already noted (using different terminology) that mental illnesses are homeostats that preserve pathology instead of resolving it—that is, mechanisms that preserve themselves (through negative feedbacks) often at the expense of an improvement in the health of their "owners".[4] Even neurosis, from this point of view, is a set of homeostatic responses aimed at protecting the stability of the barrier that separates the consciousness of the individual from its repressed content.

The metaphor of the homeostat is heuristically valuable here because, for instance, it guides the therapist's attentiveness to what is happening before him. For example, it draws his attention to the fact that the force of a neurotic patient's defense is, under normal conditions, directly proportional to his proximity to the sensitive content: in this sense, it is no different from the skin that thickens in those spots that tend to be exposed to friction, or from the body's sweating in response to heat. And this perspective, of course, naturally invites the question about the conditions under which the homeostat will become not a negative but a positive feedback.

Following Freud, and directly influenced by cybernetics, social psychologists of the mid twentieth century similarly described the various behaviors of families and groups as homeostatic systems. Here, too, the homeostat is an abstract system that appears almost like an independent entity, a system composed of the human individuals who make it up—in other words, the group dynamic is created and maintained by virtue of the integration of the various communication habits of the individuals that make up the group. In cybernetic terms, what we have here is a servomechanism made up of the response patterns—the warp and woof of the habits, sometimes pathological—of the group's individuals. These individual habits or patterns of behavior maintain one another as a delicate homeostat, sometimes at the expense of and in light of the "challenge" posed by an improvement in the health of the individuals who embody them. Thus, the recovery of the individual as individual and the stability of the group as a group often appear to be locked in a struggle for survival (this, by the way, is why the people who most inhibit our development are oftentimes those who are closest to us). Eric Berne, for example, became famous at the time for describing these group dynamics in detail as "games", that is, a set of predictable moves, or transactions, which are effectively negative feedbacks to possible deviations from a fixed group structure (Berne 1964). We can take as an example the set of predictable moves exchanged between an "alcoholic" wife and her "protective" husband (she is self-destructive, and he protects and defends her with endless dedication to avoid facing himself and in particular his sense of inferiority, fears of self-realization and independence, etc.). The wife, in Berne's description, steps up her drinking and thus the need for protection and care whenever her husband shows signs of independence and

[4]Here Freud is clearly indebted to Plato, whose dialogue *Phaedo* offers an explicit expression of the idea that within every adult lies a kind of child to whom the philosopher addresses himself directly, as to an independent entity complete with its own anxieties and patterns of behavior.

empowerment, increases the public and healthy expression of his selfish needs, etc. And conversely, if and when the wife shows "threatening" signs of recovery, such as a desire to detoxify or even success in doing so, he "perpetuates" her alcoholism, for example by making repeated references to her incorrigibly alcoholic character, putting her in stressful situations and even creating temptations. And so the symbiosis is enacted as a homeostat: its existence is sustained by the oppression or restriction of the independence of the individuals that constitute it. Similar analyses soon became prevalent also in the vast literature on work relations in large organizations, and their popularity increased the more their authors emphasized the supposed promise of such analyses to improve the efficiency of the organization, qua system, regardless of the nature and identity of its individual members (and often at their expense). This is, of course, a main appeal of studying the communication between individuals: the promise of improving the efficiency of the systems they make up.

In tandem with Freud, the fathers of modern sociology also studied human society as a homeostatic mechanism (again using a different terminology and without emphasizing the interdisciplinary similarity as convincingly as Cannon had). Thus, the functionalists, for example, from Durkheim to Malinowski, had already advocated exposing the meaning of social institutions as self-stabilizing efforts by the "social organism", sometimes at the expense of its members' liberties. Since I discuss the social phenomena in detail in the following chapters, here I wish to highlight only the immediate heuristic contribution of Cannon's basic idea. He encourages us—as sociologists—to ask the following question: *Given any social phenomenon, what does that phenomenon help stabilize? And drawing on Wiener, we add to this the further crucial question: What social development does it impede? This is the fundamental heuristic importance of the metaphor of the system. And we will note immediately that these are also exactly the two basic question that McLuhan urged his students to pose about any and every medium of communication.*

Another useful example is ecology, and in particular the study of the stability of ecological niches. The mutual dependence of the populations of predators and prey, for example, is very well known, and nowadays it is a matter of course to describe this phenomenon as a mechanism that self-stabilizes through negative feedback: a rise in the number of predators leads to a drop in the number of prey and thus to a depletion of the predator population, which translates into a rise in the number of prey, and so on indefinitely. The important point here is not the example itself (an example is never much good if it cannot be easily replaced with another example), but rather the fact that the principle that underlies it helps ecologists heuristically: it encourages them to ask, with every new phenomenon they comes across, how the phenomenon helps maintain an ecological balance, and how it challenges other aspects of this balance. And these are the lessons of general communication studies.

Darwin, as we know, had already described dynamics of this sort elaborately, and had himself already mentioned the fathers of modern economics as a source of inspiration. Malthus in particular is known to have had a decisive influence on Darwin's thought, and I suggest he ought to be regarded as the father of the concept of an "environment" and maybe also of the concept of a "system" (Sassower and Bar-Am 2014). For, Malthus described humanity as a system that faces a constant homeostatic challenge, derived from the inevitable gap between the relatively slow rate of the expansion of the means of existence and the relatively fast rate of population growth. His most famous claim—that famine, plague, and war are vital checks on a swelling human population—is an example of system-based heuristics at its best: without the systemic perspective, we could not even recognize the possibility that these harsh phenomena play a crucial and vital ecological role as feedback mechanisms.

And if we turn to economics, let us also mention Adam Smith, who described mechanisms that we now regard as quintessentially homeostatic, like the way in which goods and services find their value in the free market. By contrast, it is worth noting that clear-cut mechanisms of positive feedback are also already detectible in Smith as well as in some of his prominent critics, like Ricardo and Marx. The most conspicuous example in Smith's theory is the assumption that constant growth is possible. Another example is Marx's argument that capital gradually and necessarily accumulates in the hands of an ever-shrinking group of capitalists whose wealth becomes increasingly great (until the inevitable interclass explosion). These positive feedbacks give human history in Smith and others a distinctly linear trajectory, a direction from which there is no return—and this in contrast to the circular, equilibrium-maintaining orientation of the homeostasis and negative feedback (Sassower and Bar-Am 2014).

Cannon's great contribution, then, is in drawing everyone's attention to the crucial systemic resemblance in light of which all of this fascinating intellectual activity takes place across disciplines. This is why his basic insight rightly underlies the touching optimism of early pursuits of a simple introduction to communication. Once Wiener had articulated Cannon's insight in mathematical terms and expanded it through the notion of positive feedback, many scientists of his generation believed that cybernetics would realize humanity's (unattainable) dream of unifying the life and social sciences under the authority of a complete theory of systems.

Yet it was Wiener himself who also warned his intellectual milieu against hasty, far-reaching conclusions. He emphasized the enormous gap between successfully predicting the behavior of an ideal and fully isolated cybernetic system (such as an AC system laid out on the engineer's platonic drawing table) and successfully predicting the behavior of an actual system, like the functioning and fate of a particular air conditioner in a private home (Wiener (1948) 1961, p. 162). The problem, as we have already seen, is not just a function of the almost inconceivably complex activity that we would have to analyze to reach a satisfactory prediction; the problem is that *complexity is amplified beyond all imagination once we note that in the absence of a complete reductionist theory we have little choice but to describe the behavior of the various systems of our world as governed by mutually-adjusting feedback mechanisms: a system from one point of view is the homeostatic check on another system from a different perspective, and* vice versa. And in many cases, the understanding of the behavior of every such system is achieved within a different theoretical level, one that is not simply reducible to the other levels. Thus, the systemic perspective on our world actually highlights the need to meet the sweeping hopes of the reductionist with humility and modesty: in order to predict the behavior of even a single, simple AC system in the human world we would need first to possess a mathematical representation not only of the feedback mechanism governing its behavior in a fully isolated, ideal, setting (itself a task that cannot, as we have seen, be carried out to the full satisfaction of a strict reductionist because it is irreducibly *telic*, as all platonic representations inevitably are), but also of every possible influence on the behavior of the air conditioner, including influences that are being studied by means of other levels of description, like theories of psychology and of sociology, which to the best of our knowledge cannot be unified by means of a strict and comprehensive theory. *It follows that a pre-condition for producing the complete and accurate prediction of the behavior of a simple AC system in our non-ideal world is completion of the reductionist project, which cannot be completed…* we would have to lay out and describe a total and comprehensive history, a total and comprehensive sociology, psychology, biology, and so forth. Moreover, we would have to complete the translation of all of these disciplines into a single and complete theory. And of course the completion of each one of these supposedly distinct projects simultaneously presupposes the completion of all the other projects, because they are mutually dependent.

Above all else, it is important to reiterate that the problem here is not purely the level or degree of complexity involved: it is not just the problem of the sheer number of elements that we would have to take into account in order to predict successfully and accurately how any given system will behave. The problem is that a systemic conception of the world invites what adherents of cybernetics quickly came to call "second-order cybernetics"—that is, a description of the world in terms of various systems that function as the checks and influencing environments of other systems, alongside the fact that systemic heuristics by definition already rules out our ability to approach the problem from a purely reductionist perspective and thus to remove this mutual reciprocity by means of a strict and comprehensive theory. As we saw in previous chapters, then, the problem is both metaphysical and methodological: markedly different levels of attempts to understand the world all partake in the activity that we call "communication". And as if all that were not enough, we have not yet brought into account the most thorny case of mutual adjustment: we, the system that seeks to study its own behavior, influence and are influenced by the behavior of the systems whose behavior we seek to study, and often we influence and are influenced by the conclusions of our studies. We are all too often transformed by the conclusions of our own investigation of various levels of our existence. A complete theory of homeostasis, then, is beyond our reach for the same reason that we do not possess a complete theory of emergence or a comprehensive theory of goals: we cannot unify all contexts in light of a final and comprehensive meta-context that renders all the others redundant, at least not seriously: we cannot have a strict and comprehensive theory of communication.

We can further clarify this crucial point as an expression of what we witnessed in earlier chapters about the nature of orientation. As we observed, orientation is enabled by the narrowing-down of reality's countless aspects in light of the more or less defined goal of a given system. The reduction of orientation, therefore, just like the reduction of context, involves removing (explaining away) the goal that underlies it within a formal or extensional framework. But such a reduction necessarily fails to capture the self-orienting nature of the system it describes: the reduction portrays the system's behavior as structured and thus as known or given in advance. Thus, in the best case scenario, the reductionist is conflating a successful ex post facto description of orientation with actual orientation, that is, a theory of behavior with actual behavior, pre-knowledge with constant conjecturing, fumbling and searching, etc. This, however, is simply to forget that reduction of context is made possible only by virtue of another context (so far not reduced). Reductionists tend to recall this fact only when discussion turns to the question of whether this further embracing context can be reduced too, in turn. The answer is always positive, of course—but at a price that the strict reductionist cannot afford: it is reducible by virtue of yet another context. Our way of understanding feedback systems is therefore unavoidably multilayered and context-dependent. The hope someday to achieve a complete reduction of this context-dependence, by means of a complete and final formal uber-theory, rests on a basic misunderstanding of the prerequisites for a satisfactory reductionist explanation. Otherwise, it is merely a curious denial of the fascinating fact that communication occurs and is occurring even here and now between you and me.

Since we do not anticipate that we shall ever have a complete and final theory of communication, because "communication" is the somewhat vague name that we have given to a very large sum of emergence problems that will not be fully resolved in a single reductionist theory, it seems to me that scholars of communication need to openly part with the idea of a complete integration of their discipline into the exact sciences sometime in the future. This would entail, for instance, parting with the strict quantitative positivism that is based on pretending that this complete integration will indeed someday become a simple matter of fact. *Cybernetics helps us recognize this: it helps clarify the impossibility of a comprehensive reduction of the social sciences and the life sciences to a more basic science. At the same time, it helps us train students to recognize the benefits of cultivating high awareness to the limitations of the best reductionist models available in their field, instead of numbing their awareness with the promise that such inevitably defective models can be improved indefinitely (which is true) until it will no longer matter that they do not perfectly fit the reality that they are designed to describe (which is unknown at best, and most likely false). As part of this training, we need to sharpen their understanding of the basic difference between a deductive theory and heuristics, between a strict and comprehensive explanation and a useful but incomplete piece of advice.* Students must internalize the vital importance of a modestly goal-directed heuristics, not as a full-blown explanatory tool (this is the mistake of the eccentric vitalist or the paranoid occultist, who uses the modest idea of purpose that is imbued in so many explanations in the life and social sciences to justify what is in fact a groundless return to his own private infantile version of a world governed by spirits, by parental wills, by anthropomorphic deities and the like) but rather as a tool for exposing the limitations of the best reductionist explanation available to us at any given moment, in all areas, and especially in the life and social sciences.

Heuristics, like goal or context, is a necessarily partial and inevitably quasi-formal tool for narrowing-down the range of possibilities present before us; it is a tool that facilitates orientation where there is no complete and perfect theory to guide us (and where often we would not even want one). The importance of the role of the researcher and teacher of communication as a critic of this sort is the topic of the final part of this book. This is what I take to be the lacuna, or missing emphasis, of existing introductions to communication.

Bibliography

Arnold, D. P. (Ed.), (2014). *Traditions of systems theory*. London: Routledge.

Berne, E. (1964). *Games people play*. New York: Ballantine Books.

Cannon, W. B. (1932). *The wisdom of the body*. New York: W.W. Norton & Company.

Rosenblueth, A., & Wiener, N. (1945). The role of models in science. *Philos Sci, 12*(4), 316–321.

Rosenblueth, A., & Wiener, N. (1950). Purposeful and non-purposeful behavior. *Philos Sci, 17*, 318–326.

Rosenblueth, A., Wiener, N., & Bigelow, J. (1943). Behavior, purpose and teleology. *Philos Sci, 10*, 18–24.

Sassower, R., & Bar-Am, N. (2014). Systems heuristics and digital culture. In D. P. Arnold (Ed.), *Traditions of systems theory: major figures and developments* (pp. 277–292). London: Routledge.

Wiener, N. (1956). *I am a mathematician: The later life of a prodigy*. Cambridge, MA: MIT Press.

Wiener, N. [(1948) 2nd expanded edition: 1961]. Cybernetics: Or control and communication in the animal and machine. Cambridge, MA: MIT Press

Chapter 15
Toward a Philosophy of Communication

Before I turn to discuss the social sciences and their typical methodological quandaries, I want to make good on a promise I made at the start of this part of the book: to say a few words about the desirability of a philosophy of communication. To my mind, it should form the heart of twenty-first century philosophy. For it alone still holds the possibility that philosophy will provide both the layperson and the scientist with a vital and supporting service, as it has done for hundreds and thousands of years.

By "philosophy of communication" I mean a critical exploration of the conclusions that follow from an assumption that seems to me rather trivial and yet one that I have labored throughout this book to present as incompatible with the basic tenets of our dominant scientific picture of the world, and of course also with the bulk of the philosophical practice of recent centuries. I am referring to the (assumption of) the existence of communication right now between us, between you and me. Communication exists between us right now even if I no longer exist and you are a reader born a hundred years after my passing—itself a curious fact worthy of study. This very trivial working assumption generates fascinating problems that I wish to encourage you to explore. And it dictates a most curious position toward the traditional problems of philosophy: they become uninteresting. For with the very first theoretical move—that is, simply by virtue of recognizing the existence of communication between us—we bid goodbye to almost all the traditional philosophical problems. Clear and simple. By recognizing the very existence of communication, we find ourselves almost inadvertently replacing the preoccupation with time-honored problems that I have done my best here to explain why we stand no chance of ever solving comprehensively (because their solution requires a complete reduction of things that cannot be completely reduced) with a fascinating set of problems, urgent and enriching, whose exploration will invigorate our scientific picture of the world as well as our Ethics.

In order to see things from the point of view of the philosopher, we need to note one final time what I have tried to clarify in many different ways up to this point: *the reductionist explanation is, of course, the foundation of our scientific endeavors, and we cannot do without it, but reductionist philosophy, that is, the overall picture of the world that is founded on the promise of a complete reductionist explanation of our environment, harbors an inescapable contradiction, and one*

N. Bar-Am, *In Search of a Simple Introduction to Communication*,
DOI 10.1007/978-3-319-25625-2_15

that causes particular harm in the life and social sciences. Reductionists are right, then, when acting as consciously narrow-minded scientists or as deliberately unimaginative engineers, but they are wrong and mislead those who follow them if and when they seek to make their (consciously) narrow understanding into an overall worldview and thus into a philosophy whose goal is to provide a complete and general perspective on humanity and its place in the order of things. For, as a general, all-embracing philosophy, reductionism amounts to a promise that we know to be unfulfillable: that the best conceivable explanation of the communication taking place between us, right here right now, removes it altogether, by showing it to be a misunderstanding, indeed an elaborate illusion. This ridiculous claim is one that we should not accept. Along with communication, the strict reductionist picture of the world also removes our existence as individual entities, as humans, as well as all of our motivations as such. Accepting this loss means accepting that the very pursuit of a perspective, that is, the search for a philosophy or a science and indeed for any sort of general context that would allow us orientation in our environment, is self-defeating and pointless.

What I want to emphasize here is that, incredible as it may seem to anyone who has not paused to think about this, *nearly all the traditional problems of philosophy result from a wholesale acceptance of the reductionist effort as a general philosophy. These problems include the various well-known attempts to reduce environment (private, subjective) to reality (objective), content (context-dependent) to absolute information (context-free), active to passive, intension to extension, etc*. The most prominent of these problems are, of course, those that we have already discussed as the big problems of emergence: recall that we saw that the problem of the emergence of life from matter and the mind-body problem share the difficulty that we described as the challenge of reducing environments of orienting systems to purposeless and directionless reality. Alongside these, we encountered the problem of reducing the social sciences by means of a single explanatory theory, an issue that we take up in the next chapters and which underscores the difficulty of reducing a multilayered systemic reality to a single, comprehensive "one-dimensional" theory. But in the present context it is worth mentioning also several central philosophical problems that we have not touched on directly and which (for pretty good reasons) interest no one other than professional philosophers. These professional philosophers thus often seem like diligent sailors laboring to clean and polish up a sinking ship, or worse, to endlessly repair an imaginary ship in hope that it will carry them to the safe shores of reality. I am referring to those whose work concerns the nuts and bolts of problems such as proving the existence of other minds, demonstrating the existence of an "external" world (founding an objective reality upon the existence of private and discrete sensations, and vice versa), and even the familiar problems born of the search for a methodology that would solve all such reduction problems—most notably the problem of induction. For induction is none other than the promised method for guaranteeing the final reliability of reduction by justifying the interchangeability of intension and extension, of content and absolute information, of environment and an extensional model, etc. These problems, then, are all just different aspects of the adoption of an unrealizable ideal as a *weltanschauung* and as dictating our one and only working framework.

My point here is just this: that all these problems disappear at once, and from the very get-go, once we recognize the existence of communication as the foundation of our picture of the world, and especially as an explicit and welcome metaphysical foundation of the life and social sciences. Here is a simple example: if we accept the basic premise that the language-speaking individual is partly the product of a complex process of internalizing interactions with a language-speaking group, as the brilliant psychologist Lev Vygotsky emphasized, we will find ourselves immediately *presupposing* the existence of groups, that is, the existence of other minds, as an inseparable and trivial part of our unquestioned acceptance of our own personal existence. Solipsism (the claim that only the individual exists because nothing other than the existence of that individual can be proved, because the existence of others cannot be inferred from the individual's sensations and experiences) is then immediately revealed to be self-contradictory: *the existence of the solitary mind presupposes the existence of other minds*, so that the mind is no longer solitary. My aim, let me hasten to emphasize, is neither to produce a new version of the transcendental argument nor to establish some new sort of common-sense philosophy. And I certainly do not presume to be discovering a new America. I take for granted the fact that it is possible and indeed sometimes useful to doubt the existence of others, and of course, by all means, also the reliability of common sense (for science begins by a refutation of common-sense explanations, as already Russell emphasized). Even the existence of the self is worth calling into doubt, since we know not what the self is, and thus we can at times doubt also the existence of communication. My aim is just to clarify that I see no reason why resolving this particular doubt should be the only game in town, or indeed why this particular doubt would at all interest those who take as their fundamental problem the enigma of (the possibility of) communication. Doubting the existence of other minds is of interest mostly and perhaps only to strict Cartesian reductionists, who deny, whether wittingly or not, the existence of communication (and by extension also their own existence). Similarly, in a philosophy of communication we acknowledge upfront that the "matter" with which we come in contact is not objective reality (and cannot be that reality so long as we are individuals, i.e., so long as we are taken as orienting, and therefore goal-directed systems, indeed so long as we are taken to be alive… and when we are dead, we obviously do not come in contact with an environment since we no longer come in contact with anything at all). The scientific effort, then, is the effort to gradually expand public, or shared environments, inter-subjective environments, in an attempt to simplify and explain an ever-growing number of phenomena, to get to know them and expand them in light and by virtue of the problems of orientation that shape us. In this way, the problem of proving the existence of the external world, which is the problem of reducing contents to absolute information, is effectively abandoned at the outset. Along with it, we abandon the constant and futile attempt to establish a "logic of induction", that is, a theory that would allow us to justify our reductions once and for all.

To render my proposal more concrete (and also to avoid the impression that I am in any way presuming to invent the wheel where in fact many and greater thinkers have already done so), we can imagine a worldview of the kind proposed by Martin Buber in his famous book *I and Thou*. I prefer that we adopt this picture of the world without the superfluous theological rhetoric detailed in the final part of this classic work, and which, like every theological vestige, suffers from an anthropomorphic hyperbole that divests us of the requisite humility regarding our own place in existence. (In Buber's case, by the way, I am convinced that this rhetoric is careful lip service that he paid, consciously, in light of the boundaries of the discourse of his time, but this is not the place to expand on this point: I note it here only to emphasize that I have no doubt that Buber was aware of the far-reaching metaphysical implications of his picture of the world, which I spell out below and which he himself unfortunately did not bother to spell out because he had more pressing agendas.) Thus, we do not need an "absolute Thou" in order to recognize the existence of others and their environments, or in order that our interactions with them be taken as meaningful, just as we do not need God in order to love our neighbor or sincerely wish them good. The point I want to underscore here (and which I am somewhat bewildered that Buber did not emphasize although I have no doubt that he recognized it) is that in such a world, all traditional problems of philosophy disappear right at the outset. It is these problems that are discovered to be ephemeral.

And what is much more interesting, to my mind, is the fact that entirely new problems immediately take their place. Among them is the problem of explaining the growth of new environments in light of earlier environments and the question of the means for increasing the efficiency of communication in various contexts, that is, of the ways to improve intimacy (the ability to share problems of orientation, to share environments with others). Moral philosophy in particular stands to benefit from placing a philosophy of communication at the heart of the new philosophy. This is because, for many years, this important branch of philosophy has endured a (typically unacknowledged) problematic status whose source lies in the harmful reductionist assumption that ethics is not part of the material world and therefore in the final analysis has no place in a scientific picture of the world, except as an illusion. Many reductionists hypocritically play down or veil this assumption, but of course it cannot be escaped. Closely linked to it is the contention that science is free of any moral implications (a formal deductive system cannot stand trial for murder; an atom bomb, in itself, is innocent; etc.). It is clear by now, I hope, that these defects all result from the misapprehension that a context-free conception of the world is possible, indeed achievable, and so that our world is "essentially" devoid of human individuals, and thus of human responsibility. And it is clear, conversely, that with the opposite picture of the world, which presupposes the existence of context as a starting point, even if only as a starting problem, the (existence, or problem of the existence of) the individual quickly follows suit, and with it (*à la* Buber and Vygotsky) arrive the others, and the responsibility toward them. Responsibility to others, and toward ourselves of course, is therefore an integral part of any conception of the world that posits communication as a basic fact, or problem, and this in stark contrast to the reductionist picture of the world, which seeks, tacitly or explicitly, to do away with responsibility altogether.

For over half a century philosophy has been vacillating between two schools of thought that stand no chance of truly helping scientists and can say nothing of value to the layperson: analytic philosophy, on the one hand, and so-called postmodern or continental philosophy on the other. The first school, loyal to careful reductionist methods, has entirely forgotten the burning problems of those who seek a meaningful existence, in its attempt to articulate them as problems that lend themselves to a meticulous reductionist solution. The second school has immersed itself so passionately in the burning problems of "the human condition" that it abandoned all methods of responsible research, all inclination for rigorous method and scientific discipline, most notably the obligation that no proper study can shake to the basic structure of careful reductionist explanation. Only a philosophy based on communication, I argue, can set out from the burning problems of humans as humans, and proceed with proper caution toward a scientific picture of the world that will offer both the scientist and the layperson a helping hand, with an accompanying explanation.

Bibliography

Buber, M. (1970 [1923]). I and Thou (W. Kaufmann, Trans.). New York: Touchstone.
Vygotsky, L. S. (2012 [1934]). Thought and language. Cambridge: MIT Press.

Part III
Toward the Simple Introduction to Communication

Chapter 16
The Simple Introduction to Communication: A Methodological Preface

We turn now without further ado to the social sciences and to the discussion of what is normally described as the core of communication studies, that is, of what any decent introduction to communication ought to include. Here it is immediately clear that all the methodological difficulties that this book was dedicated to elucidating reach their glorious and terrifying climax. At first glance it may seem, mistakenly of course, as if the focus of the standard intro to communication on just one weighty problem of emergence of the three that we considered here—namely, the emergence of the social institution from the coordination of orientation among individuals—reduces the scope of our challenge. But soon enough we realize that this is not at all the case, since all the major problems of emergence contribute their share to the incredible complexity of the elusive mutual relationship between individual human beings and the institutions they establish. And as is usually the case, if we are not sufficiently aware of the difficulty, it only gets more taxing. Nothing, it seems, is less predictable than our cultural future, and thus than the future of the individuals who will be shaped by this culture. For the individuals that we are now discussing are obviously organisms, with a will, a consciousness and values that motivate them to act—that is to say, they are obviously products of the other mystifying types of emergence that we noted, as well as active contributors to their complexity. Studying individuals in their social context is therefore first and foremost *a distinctly multilayered investigation*: it necessarily encompasses central aspects of the existence of these individuals as pieces of matter with direction and meaning (cybernetic meaning, perhaps, i.e., modest but nonetheless one that presupposes a goal). Our subject matters are both propelled by their environments and actively shape them, they study their environments consciously and are transformed by the course and conclusions of such investigations. Thus, *we are never farther removed from possessing a strict and comprehensive theory than when we act as researchers of society in general and of communication in particular. After all, communication is the name that we have given to the generators of instability across all the various familiar levels of description. In other words, communication is the reason that we have no perfect definition of communication and no complete theory of communication and no final intro to communication.*

N. Bar-Am, *In Search of a Simple Introduction to Communication*,
DOI 10.1007/978-3-319-25625-2_16

So it may seem as if our simple introduction to communication ought to start with an apology that the state of the social sciences from the point of view that we are currently exploring is so complex and problematic that we cannot offer genuinely scientific theories in these disciplines. This view is incredibly common: it is a variant of the curious view that social science is not real science. I think this view is wrong and represents a dangerous tendency, since this verbal debate expresses an attempt to hide a difficulty that we should actually be highlighting as a preface to any sort of introduction to communication. For, the difficulty that we encounter here, as we have seen, is generated by the limitations of the strict reductionist picture of the world, the one that seeks to be complete and consistent, stable and infinitely accurate, and free of the limitations of individuals or of social institutions. It is ironic, therefore, that this aspiration to produce a complete and comprehensive picture of the world is precisely what drives reductionists to banish what they cannot explain from the kingdom of science. We may say that they are looking for the coin under the streetlight when it is obvious that they won't find it there because it was lost somewhere else altogether—and to this strange behavior they add the inexplicable pronouncement that if we do not find it under the streetlight, then it does not and cannot and should not exist at all.

So, for the sake of those who think the social sciences are not real sciences we should note first that *it is possible in every discipline to articulate general statements that we take to explain events. And it is possible, of course, to try to offer testable predictions that follow from these statements, and then to try our best to falsify them. These general statements are perfectly legitimate scientific descriptions, then.* And it makes no difference what phenomena they are about. Thus, *we have many and varied scientific theories of communication: their scientific nature turns on our willingness to seek their refutation and to admit it. The problem is not their status as science. The problem (if it is a problem) is that they will never be complete and flawless.* It follows, then, that *the main difficulty embodied in the simple introduction to communication is to openly admit that all of our theories of communication are fundamentally flawed, they are inaccurate, and thus flawed, from the very outset, if we regard them as strict and comprehensive reductions of communication. Even the best and most useful among them.*

Herein lies the rub. And I think that *this deserves to be the starting point of the simple introduction to communication* (indeed, I think all introductions in the social science should open on this note, and in particular introductions to communication). It ought to start with a full disclosure that is also an open statement of humility: *as far as communication is concerned, the reductionist presumption to provide a strict and comprehensive and timeless explanation of all phenomena, past and future, and in particular of those of the far distant future, reflects a misunderstanding of the objects of investigation of this particular discipline*. This is so, because articulation of any comprehensive and strict reductionist theory of communication almost always goes hand in hand with (and indeed leads to) the theory's refutation. Popper noted this point in the context of his famous critique of historicism, and made it a fundamental tenet of his method.[1] He distinguished between two kinds of historicist theories: pseudo-scientific, on the one hand, and scientific-but-false on the other. Sometimes he divided a single theory into two theories with this distinction (in some cases possibly creating unnecessary confusion by failing to draw his readers' attention to the fact that he thus handles two markedly distinct aspects of the same theoretical framework). A good example is his treatment of Marxism. Is it a scientific theory, he asked? (I set aside for now the fact that Popper did not always adequately clarify that it is not the theory itself that is scientific or non-scientific but rather our critical attitude toward it that makes it so.)[2] The Marxist claim that a final showdown between antagonistic social classes will arrive is no more scientific than the claim that the Messiah will arrive, Popper famously argued, because it is similarly irrefutable. The reason for this is that the failure of the Messiah to arrive does not itself falsify the hypothesis that he will someday come, and the same goes for Marx's final showdown. Popper adds, however, that Marx also claimed that the economic condition of the working class will deteriorate consistently with the progression of history as the direct function of a social and economic dynamic that cannot be altered. Now this hypothesis was the product of a sober, exemplary analysis that Marx conducted (*à la* Ricardo); and sure enough, it was soon refuted. Moreover, as Popper emphasizes, *the fact that Marx's analysis was perceived as an alarmingly accurate explanation of the state of affairs in his time was a major cause for the broad effort and political reorganization aimed at changing this predicted course of events*: the rise of social-democratic policy cannot be understood other than as an implicit (and at times explicit) admission of the truth of this aspect of Marxist theory, an admission that turned this Marxist prediction from a general truth into a limited truth, and thus, from the point of view of the Marxist pretension to articulate a general and strict social law, into a falsehood. Today, of course, Marx apologists say that the improvement in the condition of the proletariat was a temporary, passing phase, that Marx was right in principle but got some of the details wrong, etc., but from the theoretical perspective that we are now exploring, these lines of defense are entirely irrelevant: *the original prediction was refuted, and it was refuted precisely by virtue of having first been perceived as true*, i.e., as an expression of a generalization that admits of no exceptions—and this is the important point for our purposes, for it reflects a typical truth about social science hypotheses, and in particular those that concern communication between conscious individuals. This is the basic fact that the authors of the simple introduction to communication ought to emphasize.

There is no need in this book to enter into an analysis of the details of the various theories that make up the standard introduction to communication. That kind of discussion would necessarily draw me into a study that this book, in its capacity as

[1]See Popper (1945), Chap. 13, Popper (1963), Chap. 16, and of course also Popper (1957), where this observation is central to his attack on reckless social scientists.

[2]See, for example, the clear tension between Chaps. 1 and 16 in Popper (1963). For more on this matter and on its possible source in Popper's polemics with his positivist interlocutors, see Jarvie (2001) and Bar-Am (2014).

a meta-theoretical survey of the methodological state of affairs in the field, cannot and ought not contain: as far as possible, I would like to avoid provoking antagonism among my fellow researchers, adherents of the various approaches to communication studies. My sincere goal is to be of use to everyone. My intention, therefore, is just to point out what I think all introductions must offer their students, regardless of their school of thought. For those who are familiar with introductions to communication, and for those who plan to get to know them, I want to mention nonetheless what seems to me to deserve the close attention of all scholars of communication. Unpleasant as it may be to face this fact, we have to admit that in many cases, standard introductions to communication offer a hasty and sometimes even defensive treatment of the fact mentioned above: that from the reductionist point of view, even the best conceivable theories of communication are necessarily reductions of communication, and thus are fundamentally flawed, and in a way that should be noted as a service to students. It makes no difference, in this respect, if the theory at hand is, say, Shannon's theory, which everyone is happy to denounce as an all too limited all too formal and extensional portrayal of our social interactions, because it was not conceived by its creator as a sociological theory at all, or any mainstream theory within the social sciences, which *was* conceived as a detailed and comprehensive explanation of social phenomena.

As far as I can judge, the most common way that introductions try to play down this difficulty is to posit an empty pseudo-scientific question at their center and arrange all of their content around it. My own impression is that this strategy is common in a wide range of courses in the social sciences, and is more to blame for their unjustly dubious reputation as pseudo-sciences than the actual methodological challenges whose presentation was the business of this book. For many students, this approach evokes a vague sense of despair, whose source they cannot quite pinpoint, about the value of the theories they are learning. A prominent example of this kind of pseudo-question, a staple of introductions to mass communication, is: "To what extent do media shape the consciousness of their users?" Here students learn the various theories of communication listed in the syllabus as if they were competing conjectures about degrees of influence of various media upon their users. The discussion that develops around this question is inevitably fruitless because even a layman, let alone a critical and advanced student, immediately intuits that different environments have different effects upon different individuals and because our awareness of the environment's influence in any case alters it, as we have already noted more than once. In an attempt to skirt a straightforward admission that all strict reductions of communication are false, and in an attempt to minimize the search for any real conflicts between the various reductions, which would entail openly admitting that at least all but one of them is false and perhaps all are false, standard introductions present the various theories of communication through cases to which they apply most neatly, cases that present them at their best. Now, in itself, there is nothing wrong with such a presentation; it is important to get to know the appealing and useful sides of every theory. But not if this replaces or sidelines discussion of a theory's refutations and of the search for its points of conflict with competing theories. Often, the excuse for sidelining this vital critical discussion is the claim that the various (necessarily mistaken) theories of communication are not real theories (this is another version of the defensive view that social science is not real science). Thc theories are then described as different "approaches" to the study of social phenomena, or worse, as paradigms or as research programs. *Few tendencies are more harmful to the scientific status of a study than the tendency to blur the boundary between a theory and a research program, because the application of criticism to each of these entities is a distinctly different procedure; in particular, theories but not approaches (not research programs) can be refuted. And as we will soon see, the temptation to blur this distinction is particularly great in the social sciences because the best false theories there are often also the basis for fruitful and very useful practical approaches.*

This can be aptly demonstrated by noting the striking similarity between the common and futile debate about the extent of influence of the media upon users and the common futile debate in biology about what shapes us—nature or nurture? The obvious answer is: it depends. Depends on what we are examining, on the given context, and on the level of detail in which it is described. After all, biologists all agree that organisms carry and pass on to their offspring different building instructions, some of which are translated, under proper conditions, into vital guidelines for behavior, and they agree also that without constant interaction with various environments, these instructions not only do not properly develop but are virtually useless, since without the context of the environment they are as meaningless as all pieces of information must be when taken out of context. What is often overlooked, then, is that this trivial observation pulls the rug from under the popular debate about the exact role of nature and nurture in determining our conduct—and it does so in one fell swoop, undermining at the same time numerous similarly flawed follow-up questions, such as "What is the relative weight of the influence of nature and of nurture upon our behavior?". For clearly, in certain ideal contexts, the role of heredity is more pronounced precisely because we are deliberately ignoring the idealization, that is to say, because we are consciously ignoring the fact that any fixed concept of "environment" (or "nurture") is the product of our conscious suppression of the fact that the environment can also be very different, or of our deliberate working assumption that no significant change in the environment is possible, an assumption that allows us in the first place to view hereditary matter as distinct from its environments and thus as having its own distinct and stable meaning, like a word in a dictionary (for clearly, the physicists qua physicists sees neither the word nor the genetic instruction, let alone their meaning). We need no intricate and detailed discussion of epigenetics, of hereditary mechanisms external to DNA or of multilayered natural selection environments to realize that performing an idealization of the kind that distinguishes hereditary matter from its various environments and then ignoring the fact that this distinction is the product of an idealization is about as resourceful as the Baron Munchhausen lifting himself out of the mud by pulling his own hair. We have already seen that without purposes and an environment, the very notion of "system" is meaningless. Biologists, at least, do not retreat from the difficulty we articulated here by contending falsely that there are no scientific disagreements in their field but only different "approaches" to the study of organisms… For there is, of course, a fascinating and important question that concerns the geneticist and the evolutionary biologist alike: the question is, "In what cases, satisfactorily defined, does our control over the organism's various development environments allow us to predict its behavior, in light of what we know about its past? *In this respect, communication scholars can learn from biologists, who are recently discovering with great and justified enthusiasm that a constant improvement of our ability to predict the behavior and development of organisms proceeds via the integration of descriptions of their behavior form several different and distinct descriptive levels, and thus also of distinct descriptions of their various environments (genetic, epigenetic, behavioral-cultural, and so forth—of course, in light of various theories that are described with sufficient accuracy within these levels of description…).*[3] *Accordingly the question that should concern scholars of communication and which deserves to be placed at the center of the introduction to communication is almost identical: how does our command of the various environments in which individuals can be situated improve our ability to predict their behavior? By now it should be obvious to us that this improvement necessarily entails a distinctly multilayered description of the individual's behavior and environments, and this, again, is the single most important basic fact that the simple introduction to communication ought to emphasize.* Alongside this basic fact it bears repeating also that we would do well to suppose that even our finest answers to this question will properly be refuted at some point, if not immediately upon their articulation, for the reasons we elaborated in this book, i.e., because they are strict reductions of communication.

What ought to interest scholars of communication as such—that is, as scientists—is not the successes of their theories but their theoretical limitations: they will benefit from focusing on exposing those places in which, if and when the theory is a strict reduction, it shows up as a false, over-simplified description of a reality that is more complex and dynamic than the theory at hand allows us to describe. In addition, they should note that it is precisely these refutations that will later transform the theory into the foundation of a productive research program, i.e., into effective technology in the hands of communication professionals. This last point, whose importance to those who seek the simple introduction to communication cannot be overstated, is hard to explain without saying something more general about the very confusing line between science and technology, a line that is particularly fine in the social sciences and especially in communication studies, where it marks *the all-important difference between the communication scientist, on the one hand, and the professional practitioner of communication on the other hand.*

Joseph Agassi was the first to articulate what is now accepted as the basic difference between science and technology: the respective attitudes of those who work in these distinct domains toward refuted theories (Agassi 1985). Agassi corrected the traditional mistake of viewing technology as an application of the

[3] Jablonka and Lamb (2005) is an excellent recent example of this trend.

latest scientific knowledge. This is a mistake, he showed, first, because quite a few technologies are simply not based on science at all, let alone its latest version. Second, and more importantly for our purposes, *the vast majority of technologies that currently serve us were designed in light of scientific theories that are already known to be false*. A prominent example is Newton's theory, which has of course been refuted, and underlies more practical technologies than any other scientific theory that I can think of. Indisputably, its usefulness to the engineer is greater than that of the theory of relativity, which replaced it: we made it even to the moon based on Newton's theory—many years after it had already been widely acknowledged as an incorrect description of reality from a strict and uncompromising reductionist perspective. *The scientist, then, insofar as she is a strict and steadfast seeker of the truth, abandons the refuted theory when it is refuted and because it is refuted. And insofar as she is a strict reductionist, she is of course never fully satisfied with a theory that is merely an approximation. For the media expert and the communication engineer, by contrast, the refuted theory, precisely when it is considered as an approximation, is typically easier to apply.* This is undoubtedly the case if we compare Newton's theory to Einstein's theory, but it is also true more generally, because the way to show a theory to be a mere approximation and to improve it is typically to improve the resolution and accuracy of its description, and this in turn almost always increases the complexity of the description of phenomena significantly and sometimes even requires a new level of description. Therefore, strange as it may sound, *it is easier to think and plan our moves with the help of the falsified theory. Moreover, refutation allows the technician and the engineer to know the precise limits of the successful application of the technology that they are using.* The limits of its application are clearer than they are in cases of technologies derived from the theory that has not yet been refuted and which may someday be refuted in ways and in cases that have not yet been tested or even imagined.

And this is precisely the case with the best false theories we possess in social science in general and in communication studies in particular. The fact that they are false, that they are at best rough approximations of a much more complex and dynamic multilayered reality, is a point that should be openly admitted at the outset. First of all, it is an important point for the scientist or scholar because it is the motivation and the key to constantly improving and updating these theories. And in addition, it is no less important to the media professional, the engineer and communication expert as such because his main interest is in doing and applying, and his knowledge of the limitations of the falsified theory, as we just noted, makes it *more* rather than less useful to him than its more up-to-date, accurate and sophisticated substitutes. *For students in particular it is easier to study a false theory when their attention is explicitly drawn to the fact that it is false. This is especially true when we not only point out to them the refutations of each and every theory but also train them to recognize these as the basis for each theory's potential heuristic advantages.* This, I think, is our task as teachers of communication, and it is not a simple task. Recall again that from this perspective there is no difference between non-social theories of communication, like Shannon's mathematical theory, and the most advanced theories of sociology and psychology, provided that they are articulated within a strict reductionist framework. Those who present Shannon's theory only to contrast it later on with these other theories, which are cast as successful theoretical treatments of the complex activity of producing and coordinating meanings, are simply wrong and misleading, and thus do a disservice to students. Instead, they should explain that the power of all those theories that deserve to be featured in the simple introduction to communication lies precisely in their nature as extremely useful yet consciously problematic generalizations.

This book, as I have said, is written as an introduction to the introduction to communication studies, and not as another competitor in the market of straightforward introductions to the field. Confusing these two tasks would undermine my goal, which is to be read alongside the field's popular introductions and as a friendly complement to them. Yet it would be hard for me to sum up my words to students without some sort of demonstration, even very brief, of the way in which the spirit of the theoretical conclusions offered here should be expressed in the discussions typically featured in proper introductions to communication. To this end, I suggest that we sum up this book with a symbolic discussion of two examples. The first example, presented in the next chapter, concerns the study of orientation. The second, to which I devote the final chapter, concerns communication in the most narrow and widely accepted sense of the term as we have learnt to apply it here, that is, the collective coordination of orientation challenges.

Bibliography

Agassi, J. (1985). *Technology: Philosophical and social aspects*. Dordrecht: Kluwer.

Bar-Am, N. (2014). Thomas S. Kuhn's the structure of scientific revolutions 50th anniversary edition. *Philosophy of the Social Sciences, 44*(5), 688–701.

Jablonka, E., & Lamb, M. (2005). *Evolution in four dimensions*. Cambridge: MIT Press.

Jarvie, I. C. (2001). *The republic of science*. Amsterdam: Rodopi.

Popper, K. (1945). *The open society and its enemies*, Vol. 2. London: Routledge.

Popper, K. (1957). *The poverty of historicism*. London: Routledge & Kegan Paul.

Popper, K. (1963). *Conjectures and refutations*. London: Routledge & Kegan Paul.

Chapter 17
Example #1: The Classroom

Before launching into our example, we need first to introduce the two most prominent and influential reductionist frameworks in social science: psychologism, and collectivism. Most social science theories are significantly indebted to (at least) one of these schools of reduction.

Psychologism aims to reduce all social theories to psychology, that is, to explain all social phenomena based strictly on assumptions about the motivations of individuals (their values, hopes, needs, passions, etc.) as individuals and about their physical limitations as individuals. The backdrop to this influential trend is, of course, individualist liberal philosophy, which regards the individual as a rational and autonomous creature and aims to establish a society that reinforces these qualities (and otherwise loses its right to exist).

Collectivism, by contrast, is the attempt to reduce all social theories (including individual psychology) to distinctly sociological theories—theories that base the adequate scientific explanation in social science upon social institutions that serve public needs and social dynamics, which, according to the collectivist position, shape the motives and aspirations of their individual members and thus subtly dictate their behavior. Typically, collectivist theories assume the existence of social entities (whatever may be the meaning of this typically vague term), like economic classes, peoples, and the like, as well as the existence of distinct motivations on the part of these entities (whatever may be the meaning of *that* typically vague term), like a "national will", "human spirit", "class interest", etc. More importantly, however, the collectivist contends that the social entities are prior to the individuals not only from an ontological point of view (whose meaning, as we said, is all too often very vague) but also from a historic point of view and from a logical, theoretical point of view. *Human psychology, collectivists argue, is the effect rather than the cause of the cultural framework that shapes it.* And as Ernest Gellner emphasized, unlike psychologism, which grew out of the individualistic trends of the modern era, the roots of collectivist theories are pre-philosophical (Gellner 1998, pp. 7–13): our historical existence as a preliterate tribe far preceded our conscious and distinct existence as individuals. Individualism, therefore, is of course a late revolution in the annals of humanity, since it is the result of the maturing of an advanced awareness to an advanced stage in our evolution. Consequently, as a distinct philosophy and scientific method of investigation,

N. Bar-Am, *In Search of a Simple Introduction to Communication*,
DOI 10.1007/978-3-319-25625-2_17

collectivist theory emerged only in response to individualism and after it—that is, it was born as a critique of the psychologistic reductionist picture of the world.

By way of introducing the examples I discuss below, and as a service to those students who are unfamiliar with the literature, let us further note that, rightly or not, it is customary to regard the fathers of modern sociology—Max Weber and Émile Durkheim—as distinct representatives of these opposing schools of reduction: Weber is perceived as representing the individualistic-psychologistic strain of thought in sociology, and Durkheim as representing the collectivist explanation in sociology (the accuracy of these labels, or more precisely, the difficulty involved in determining their accuracy, is addressed in the closing paragraphs of this chapter).

Unlike both Weber and Durkheim, Georg Simmel is typically known as the lauded and ridiculed outsider of social science. This is not just because he never held a proper academic position (a fact that can easily be put down to his extraordinary brilliance as a researcher; few thinkers have seen as many imitations and straight-out plagiarisms of their original ideas as Simmel, and there is no surer way of getting away with stealing than keeping the victim of the theft off the main stage), but mostly because he avoided presenting a reductionist theory that is one-dimensional and easy to operate (and therefore clearly false) of the kind commonly attributed to the founders of the field (Weber and Durkheim). In other words, the explanations Simmel offered of the cultural phenomena he studied were deliberately and explicitly multilayered, and therefore also free in spirit of the attempt (and of the hope) to reduce these phenomena to a single traditional level of description. Simmel was opposed in particular to the attempt to reduce the behavior of individuals to nothing more than their psychological motivations as isolated individuals or conversely to nothing more than the institutional circumstances that shape them, and the rejection of this dichotomy became a basic tenet of his method. Of course, his thought was still based on the basic desire to explain, and so to reduce, the action of individuals to the context in which they function, but in his account, this context is multilayered and more flexible (or rather, it would be described as "multilayered" from a traditional reductionist perspective). And it typically involves a meeting and/or conflict of different forces belonging to distinct levels of description. Thus, the Simmelian discussion typically revolves around the elusive role of the tension between these two forces (the mental and the social) as the mechanism of change that characterizes what takes place before our eyes. This is why, in my view, his analyses are more accurate from the purely theoretical point of view. Now, we might have expected them to be also less simple and useful from a purely practical point of view, since we have already observed that the more accurate a theory, the less useful it tends to be. And indeed, a crude reduction is undoubtedly more easily applicable than a complex, multilayered one. Yet Simmel—and many argue that this is where his greatness lies—chose his examples with exceptional insight: he focused precisely on cases in which a multilayered description offers us a way to solve a difficulty that a simpler and more crude reduction cannot even properly describe, especially if carried out with adequate rigor and exactness. In other words, he focused on social phenomena whose explanation invites complexity as a tool for solving a problem that a simpler reduction simply fails to recognize. And so it happens that his descriptions are often preferable even from certain practical perspectives, despite their multilayeredness.[1]

At this point, in an attempt to demonstrate the respective advantages and disadvantages of the theoretical reductions articulated from within collectivism and psychologism, if only in their strictest, ideal formulation, and more specifically in an attempt to understand the very distinct importance of these reductions to the scholar of communication on the one hand and to the media professional on the other, I usually suggest to my students that we focus our attention on a typical social-science explanation. This explanation will be based on a very simple and straightforward reduction and yet one that is articulated rigorously, and which is therefore plainly false. Nonetheless, my hope is that it is still helpful and useful enough to demonstrate the problem at hand: namely, *how best to cultivate the awareness of our students as communication researchers and media professionals to the usefulness of the false theories that we teach them, without pretending that they are true*. To this end, I usually focus on a phenomenon that my students partake in without even budging from their regular seats in the lecture hall (indeed, especially when they do not budge from those seats).

An immediate example that occurs to me as my students and I file into the lecture hall has to do with our meeting with this space: the gathering space is, of course, a highly present element in the experience of those who gather in it. When I enter lecture halls as a lecturer, I oftentimes recall my visits to ancient Greek theatres and note the marked difference between those spaces and the one in which I am now in. The former embody a truly exhilarating awareness on the part of their planners to the quality of the experience (originally religious) that those buildings were inspired by, served to enhance, and attempted to translocate into other contexts. I mention this fact also in order to illustrate Agassi's aforementioned claim that in many cases—the study of communication being a prominent example—*technology precedes the scientific theory*, the latter fumbling about in a retroactive attempt to unlock the secret of the former; this, in contrast to the widespread and often false view of technology as the application of cutting-edge scientific knowledge. The pioneers of communication, past and present, inevitably grope their way about in the dark, and the paths they take are often understood only in retrospect with the help of the scientist. (Linguistics, for example, searches with evident difficulty for the keys to the virtuosic use of language by native speakers, and the researcher of body language searches no less humblingly for the secrets of our remarkably smart and typically intuitive use of our bodies; it is no surprise, therefore, that reductionist linguists, like Chomsky for example, fail over and over again in their descriptions of the ultimate reduction of all grammars to a pre-structured theory of grammar, and that the sages of "body language" learned long ago that our use of our bodies is not a language at all but rather a multilayered and sometimes dynamic alloy of quasi-flexible patterns of behavior.)

---

[1]For classic examples to such cases see Simmel on "Fashion", "The metropolis and mental life" and "Subordination and personal fulfillment" in Levine (1971), part V.

And so, to a large extent, the modern lecture hall is a late capitalist incarnation of that Greek theatre space, with everything that this implies about the radical differences in the respective experiences they provide their visitors. A well-known example is the exceptional acoustics for which the ancient Greek theatre is renowned. It may be surprising to discover just how badly the modern lecture hall fails (and indeed, wisely, does not even try) to compete with its Greek predecessor. Naturally, the quality of the intimacy that speakers and their audiences can achieve in the modern classroom very often suffers as a result of this difference. And it is clear, I hope—as this is the important fact for our purposes—that the consistent collectivist is committed to seeing the space of the modern classroom, like that of the ancient Greek theatre, as a causal factor that shapes the consciousness of those who gather in it, and nothing more. He has to study the classroom space as an expression of the particular collective unit of existence that it represents and nothing more (ignoring the fact that this unit is made up of individuals with independent wills). The consistent individualist, meanwhile, is committed to seeing it rather as the product of the aspirations of individuals and nothing more—for example, of the intentions and motives of its architects, as individuals—because for the consistent psychologistic researcher, the collective is nothing more than a sum of individuals. He studies the class, and at the same time wants to ignore the fact that it is a class (that is, a social institution).

When I teach these competing reductions I suggest to my students that we try together to explore the limitations that each of them imposes on the description of any given phenomenon that involves orientation. Together, we list different qualities of the lecture hall and make various reductionist conjectures that explain the influences of these qualities or their possible causes. We notice, for instance, the size and the shape of the classroom, the number of seats and their arrangement and degree of crowding, the quality of the lighting and the color of the walls, the height of the speaker's stage or podium relative to the eyes of the viewers and its distance from them, and other similar details. I try to devote an extensive discussion to each one of these aspects, for they are all elements that have a rhetorical presence whose influences and causes interest future communication experts. But how does the scientist of communication approach their investigation? And how, by contrast, does the media professional enhance his awareness of their uses? And how shall we, here, sharpen the all-important distinction between these two questions?

A charismatic speaker could go on at length about each one of the classroom parameters we listed, punctuating her presentation with a succession of fascinating and amusing anecdotes. This often happens in the classroom of the simple introduction to communication. But the simple introduction to communication cannot be allowed to boil down to the skillful entertainment of an audience with interesting examples. What we are now seeking is a testable scientific hypothesis that explains the conduct (or the factors that shape the conduct) of students in the lecture hall. The factor of classroom overcrowding in particular draws my attention because it has been addressed by almost everyone who wrote about the rise and characteristics of the modern era: crowding features in all the canonical analyses about the rise of the urban lifestyle.

As a first step, I ask my students to try to imagine what happens to the sense of self of each and every one of us inside a jam-packed elevator: we automatically pull ourselves upright to stand "at attention" and for a brief moment try as best we can to minimize if not altogether cancel our physical and mental presence as individuals. Various aspects of this behavior happen of their own accord, without any conscious intention on our part. Almost involuntarily, we avoid behaviors that we naturally and unthinkingly allow ourselves when we are alone at home or when we observe others from a safe distance, for instance out on the street. Even this simple and short-term shrinking of the self has immediate and evident biological, psychological and social manifestations. For example, in the overcrowded elevator we all but automatically avoid eye contact, which we might have sought out from a somewhat greater distance. And even if we do not avoid it when our faces are crammed up against each other in the elevator, our behavior will betray our heightened awareness to the fact that this close eye contact constitutes a significant invasion of our own and of the other's personal space. Any biologist can tell you that closeness and intimacy of course also bring with them a greater risk. This is why courtship, for instance, is such a protracted, ceremonial and gradual activity in the animal world: it is a mutual adjustment of permissions whose attendant risk becomes increasingly greater up to the point of great intimacy. In the overcrowded elevator, for example, we will even try to moderate our breathing. So, even from this kind of pre-scientific, intuitive investigation, we can easily observe that crowding and individuality maintain a tense relationship that can later on be sought and found in every corner of modern life: in prisons, military camps, urban slums, or by contrast in the vast private yards of the upscale suburb, and even in the distant and secluded open spaces of the mountain and the desert which attract those individuals who seek the ultimate mystical expansion (always so paradoxical) of their selfhood.

The incredible ease of this pleasant stringing of anecdotes is misleading. It trips up many students, teachers and researchers, because an interesting string of examples is so easily confused with a theoretical investigation of their meaning, which is also why the work of the scholar of communication is so easily confused with that of the media professional. My own impression is that this confusion is very typical of the standard introduction to communication. In order to discuss the possible theoretical meaning of the various examples, we need to turn away from the examples and back to the two basic frameworks of reduction introduced above, psychologism and collectivism, and to note the following important fact, which seems to me to carry staggering implications: for the consistent adherent of psychologism, the overcrowding is (no more than!) a result of the values and aspirations of individuals, as individuals. Perhaps these overcrowded individuals crowded together because of some more important personal goal, like getting to work on time or inexpensively, or finding a job, etc.; and perhaps the crowding is a tool in the hands of some more powerful individual who wants to use it to help maximize his profits. But in any case, overcrowding cannot be regarded here as an explanation of the individual's behavior, as its cause. Because in the final analysis, for the consistent adherent of psychologism, it is the individual who chooses his behavior in light of his goals, values, etc. Yet for the consistent collectivist, by contrast, the crowding is (no more than!) a property of the social entity, and as such, it alone is a causal factor that shapes the psyche of the crowding individuals; the crowding is a means by which the social institution controls its parts—for example, a means of improving the efficiency of the various parts of the system as a system. It is a tool for controlling their behavior.

(At this point it seems natural to ask: Is the fact of the existence of some sort of influence of the crowding upon the crowded individuals compatible with the psychologist assumption about their freedom to behave as they choose? Is the fact of the existence of some sort of behavioral freedom of the individual as individual compatible with a consistent collectivist explanation regarding the influence of the context on those who are trapped within it? And *is not the very existence of orientation already fundamentally opposed to the pure version of both of these proposed reductions*? After all, as Simmel repeatedly argued, if we allow for the existence of a certain contextual influence alongside a certain freedom of action, this alone implies that both reductions are strictly speaking mistaken from the very get-go.... We will get back to these questions shortly.)

Before we return to these basic questions, I would like us to try to complete our example. We need to give an example of a psychologistic hypothesis and of a parallel collectivist hypothesis about orientation, and to suggest possible experiments for testing them. So, for example, we could seek to study the cause (or influence) of the distance between the different students in the classroom or the cause (or influence) of the distance between the students and the lecturer. Let us focus on the second option, since this distance is chosen anew (or succumbed to) by the students with every class they attend—of course, within the physical limits of the given classroom. We could inquire, for instance: What is the meaning of their seating close to the lecturer or farther away from him? The planners of the earliest theatres already knew that the front rows afford a more encompassing, richer, and more significant viewing experience, which is why even in ancient times these seats were typically reserved for important members of the community and for the judging panel. Today, as we all know, these front-row theatre seats are sold at higher prices. These facts suggest a principle, namely, that there exists a certain bias toward being close to the events: proximity allows for a more complete, lively, and intimate experience of what takes place on stage. In order to complete the example, then, I propose two testable hypotheses (which I reject out of hand, of course), whose distinguishing feature we will try together to identify.

The first hypothesis is distinctly psychologistic: it states that the front rows in my classroom will typically be occupied by students whose urge for a richer, more intimate experience of the contents of my lecture is greater. Usually, these students can also be correlated with a broader urge to excel in their studies, particularly when they manifest a consistent choice of front-row seating across different classes and lecturers. Thus, we can raise the psychologistic hypothesis that students who sit in the front rows are motivated by an ambition to excel that is stronger than that of their back-row peers. (Of course, we will also find other—or potentially other—cases in the front rows, like short-sighted students, special guests, students with a special interest in the lecture topic, and many other cases that I will not analyze in detail here because my point is not to give a satisfactory explanation of my students' behavior but to illustrate a consistent reduction, false but useful, of their behavior to one of the basic explanatory schemes of social science. Having said that, let us note that it is certainly possible to try to bring together all these supposedly exceptional cases under the same reductionist framework of "a personal desire to enhance the lecture experience", whether the cause of this desire is organic, as in the case of short-sightedness or ADHD, or other, and then to reduce these various other desires to the desire to excel.)

Now, as a complement to this psychologistic hypothesis we can add the hypothesis that the back rows are chosen by those students who prefer something else over doing well in class—for example, the freedom to ignore the lecture and preoccupy themselves with something else, or maybe outwardly exhibiting their rejection of the value placed on academic excellence by front-row students, or exhibiting an ambivalence about belonging to a group with the rest of their classmates (or else exhibiting their desire to belong to this group but specifically as marginal members), and so forth. Here, too, of course, there will be various other cases, whose status as real exceptions or merely apparent exceptions (reducible to the norm) can be considered later on. A good example is the case of late-comers: typically, late students go to the back rows; so perhaps they are just a sub-group of the ambivalent students… or maybe not. And so forth.

Next we need to propose a collectivist research hypothesis concerning the same state of affairs, and then to ask whether these two are competing hypotheses. The collectivist hypothesis is of course a mirror image of the psychologistic hypothesis, for as we noted, where the adherent of psychologism sees a cause that drives individuals to action, the collectivist sees the forced movement of individuals propelled by a systemic structure and serving a systemic purpose. The collectivist hypothesis, then, is that a random student placed consistently in the front row will gradually do better in class than when she is placed in the back row, and will, moreover, gradually find greater interest and take greater pleasure in the studied topic, and even—most importantly, perhaps—will gradually come to express the "spirit of the class", without which, the collectivist contends, the class is not a class. According to the collectivist, physical proximity to a lecturer shapes excellence, regardless entirely of the student's original abilities and ambitions and original level of interest in the material. My students tend to be far more impressed by this hypothesis than by the psychologistic hypothesis, revealing something of the enormous success of collectivist theories, and in particular of neo-Marxism, in communication courses: collectivism is so successful because it places an illuminating emphasis on the impact of *the hidden context*, the context that escapes our eyes. In the case of our example, this context is a rather trivial fact but one that is unfortunately often overlooked: the designers of the modern lecture hall have accepted without question the assumption that it is not possible for all the students to sit in the front row. This assumption is disturbing because if it is indeed true that a growing distance from the lecturer is ultimately expressed as a greater learning challenge, then large-scale classes with a fixed seating arrangement more or less guarantee poor students! Although the seating is of course not set in stone or fixed by law, still, in the interest of energy preservation, self-orienting organisms all tend to form fixed patterns of orientation and movement. Thus, it is very rare for my students not to seek fixed seating in any environment to which they return regularly. The significance of all of this is rather incredible: even if we are somehow able to make several clones of a given "straight-A" student, if we make a point of assigning them regular seating in all different parts of the classroom, we will soon find ourselves with a range of grades that is typical of any reasonably heterogeneous class and with a similar range of levels of student interest, just as if we had planted a set of genetically-identical seedlings in very different habitats. A collectivist explanation of the inevitable existence of "bad seats" will therefore also typically include a comprehensive discussion of the question of why weak students are useful and indeed vital to the society of which they are members, for example because of their contribution to class morale, to the economy, etc., and similarly a discussion of the usefulness of crowding to the social organism.

Now we need to ask whether our research hypotheses can be tested empirically. It seems as if we need no more than to design the right experiment. To test our psychologistic research hypothesis, for instance, it is commonly assumed that we need simply to propose a set of testable parameters of a personal ambition to excel, for example the kind of parameters that can be assessed by asking students to fill out questionnaires, or by interviewing them, etc. Then we rank them according to their level of ambition, and based on this ranking, produce a prediction about their approximated location in the class. Finally, we test the verity of our predictions.

And what about the collectivist hypothesis? Here it seems that what we need is simply to ask students with similar intellectual attributes and similar patterns of achievement to sit (consistently) in different places in a large lecture hall, and after a certain period of time, to assess the changes in their academic achievements, or perhaps their respective levels of satisfaction with the lectures—whichever we prefer.

Before I am suspected of excessive naïveté, I ought to stress again a point that I have emphasized repeatedly throughout this chapter: the hypotheses I raised here about (the effect or the cause of) the distance from the lecturer are plainly and deliberately simplistic, and thus are true at most as rough approximations of a more detailed description of what takes place in the classroom. Thus, they are obviously false from a more rigorous and strict perspective. *They are false because they are overambitious reductions*; this, indeed, has been my argument throughout the book. But what I tried to show with these hypotheses is that *the main challenge for the authors of the standard introduction to communication* (in the narrow sense of the word) *is* not the lack of theories, simplistic or complex, more or less detailed, but *a clear and unapologetic presentation of the fact that they are all known in advance to be false reductions.*

This understanding in turn allows us to *highlight the line between scientific research and its applications* rather than blurring it as all too often happens in communication studies. For instance, note how even without any rigorous empirical research, *the mere fact that I was careful here to articulate the reductionist hypotheses as strict reductions, immediately exposed their excessive rigidity*:

human orientation in the environment is a partly initiated action toward a goal within the context/environment in which it takes place. Thus, if the context/environment alone explained our behavior, as the collectivist argues, we could not be perceived as active initiators of this behavior and the explanation of our behavior could not include any explicit reference to our goals… goals would then be presented as illusions that accompany our blind obedience as detailed in the strict collectivist explanation. On the other hand, if and when our action is explained as the result of nothing more than our initiative, as the adherent of psychologism proposes, that is, as the result of the motivations of individuals as such without acknowledging the ways in which these motivations are shaped by the context/environment, then we have in effect described ourselves as utterly free of any influence by our environment. But *anyone who is completely unaffected by his environment* (*and it makes no difference in this case if the environment is external or internal*) *is not in the environment in any sense at all*! If a totally free individual existed she could not even notice her social context. Like the omniscient God, she is no longer an individual. When we assume that it is possible to ignore the environment altogether, for instance by supposing that the individuals before us are perfectly rational, or when we assume, conversely, that their nature as individuals is a negligible attribute, we are ignoring the fact that we are dealing with self-orienting systems. And when we do that, we are no longer studying communication in any accepted sense of that word. Another way to put this Simmelian point is this: *every explanation of behavior qua behavior is necessarily multilayered and incomplete*.

(And indeed, when my students hear me utter a certain word—for instance, "cat"—there is no doubt that if they understand my language they will have no real choice but to picture a cat in their minds… this momentary conjuring of a mental image is an almost mechanical response by language-speakers to utterances articulated in their language and spoken in their vicinity. But when I tell my students that there is a kitten in my shirt pocket, even though they cannot resist being activated at the level of their understanding of my words, still, most of them deny the truth of what I have said. In other words, they are activated on one level and are considerably autonomous on another level. *This multilayeredness is the heart of every credible description of the act of communication, and thus it must become the recognized feature of any theory of communication that does not want to wipe out of existence its own object of investigation*. The ability to understand what is said *depends* on the loosely hierarchical and simultaneous existence of contexts, and these contexts must allow for flexible activation on various distinct levels.)

Thus far, we presented two research hypotheses. Both of them contain a certain limited portion of truth (which is to say that both of them are strictly speaking false). We now need to focus on a particularly disturbing consequence of our discussion that has severe methodological implications. In order to achieve this we should note first that as strict generalizations, *the two competing reductions we presented here were constructed so as to cancel one another out*: *they were designed to be competing descriptions of the same conduct, in the same classroom*. Yet rarely are they presented as such. The question we need to address, therefore, is not whether we can refute these hypotheses: for example, whether we are able to find "straight-A" students who choose consistently to sit in the back row or students whose grades will rise precisely because they have been placed in the back row (of course there are such students). (A more interesting but for our present purposes still marginal fact is that students who are exposed to the conclusions of our proposed study may elect, in light of these conclusions, to change their usual seating in the classroom, thereby rendering our hypotheses false even if they once were true.) The question we need to address is how to handle the fact that each of the two explanations we described above made an indispensable contribution to our understanding of events: *they are thus complementary as well as competing*! *In particular, the facts as described in our supposedly competing hypotheses support and enhance one another. This occurs despite the fact that one theory describes a fact as a cause and the other describes the same fact as an effect*. Thus, even if the subjects of our experiment were never exposed to its conclusions, we would still be likely to find ourselves at once affirming both (contradictory) hypotheses with our experiments on the same single class! Now, of course, from a purely methodological perspective, this outcome is completely unacceptable, since the object of these opposing reductions is to cancel one another out and not to complement each other. The fact that we have not succeeded in rendering either level of description redundant means that we are unable even to articulate the distinction between the desire to excel as a cause and as an effect.... It follows, sadly, that we have failed to provide an adequate explanation of what takes place in the classroom. And communication is the source of our conundrum.

The methodological conundrum we face here is thrown into sharp relief thanks precisely to our strict adherence to the rules of reduction in the face of a phenomenon that does not lend itself to strict reduction—namely, orientation. It is therefore important to observe that the reductionist's most common defense against publicly acknowledging this conundrum is the one described throughout this book as the typical reductionist response to any falsification: the retroactive correction (ad hoc). Now, this move, as we already discussed, represents welcome scientific progress when it is performed openly and with full awareness, i.e., when the correction is public; revising theories in light of falsifications is after all the heart of the scientific practice. *The problem arises when the correction of the psychologistic theory is achieved through a silent, retroactive incorporation of collectivist facts into the psychology, and vice versa. I think that this procedure is the main reason for the bad reputation of social science among the sciences.* For it is this strategy that blurs the clear line between the different levels of description. Now, even this blurring is not objectionable in principle, so long as it is done openly and honestly. The psychologistic researcher, for instance, might accept as a fact the collectivist contention that the structure of the classroom affects the character of those who study in it, and even that this effect happens in light of the needs of society in general, while still describing this state of affairs and these general needs as the product simply of a sum of aspirations of individuals as such. This move "saves" the psychologistic theory from being falsified but at the high cost of making it harder to draw any meaningful distinction between a psychologistic and a collectivist explanation, since the adherent of psychologism hereby effectively admits that a purely psychological explanation is necessarily incomplete. The collectivist, by contrast, might accept that there are cases in which the behavior of certain individuals goes against the trend that drives the collective as a whole, and even against the direction that is most desirable for the collective as a whole, while still trying to explain this deviant behavior as the product of some further collective aspiration that has not yet been brought into account and which would show the "deviant" behavior to be merely apparently-deviant. Thus, *instead of two competing trends of reduction, we find ourselves gradually in the midst of a tacit process of a constant blurring of the distinct levels of description....*[2] This methodological defect is unnecessary, and it is responsible for making many of the social science theories *at once false and unfalsifiable* (exactly as Popper described them). Popper's claim initially appears to be insupportable because at first glance it appears that any theory that cannot be falsified cannot be false. In fact, however, its falsity becomes apparent when the scientist proceeds with caution, preciseness, and clarity, and it is irrefutable in principle only when the scientist adopts the habit of re-formulating his theory ad hoc in the face of falsifications so that its new formulation incorporates the falsification—in other words, when the scientist blurs the line between the reductionist theory he currently possesses and the reductionist research program to make it, someday, into the final true theory. *For the theory was and remains false as a theory so long as it is articulated with sufficient clarity and detail, and it is irrefutable in principle only as an ultimate research plan. This blurring of the boundary between theory and research program, I argue, is the main reason that we almost never find straightforward, consistent reductionist explanations in introductions to communication: presenting them in this way almost automatically reveals them to be plainly false. In an attempt to avoid this consequence, authors present them as broad research plans. As such, however, they are in principle irrefutable, and thus their scientific status is undermined.* And this is how the all-important difference between research and technology comes to be obscured in the field of communication studies.

We are left now with one final question, namely whether and how all this helps our practical-minded students? Most of them, after all, do not want to join the research community but rather want to become media professionals—journalists, marketers, etc. My contention is that our ability to be of use to them turns on our ability to help them draw a clear distinction between their training as researchers of communication and their training as media professionals. Moreover, the professional learns more easily by observing the researcher (so that we should encourage collaboration between these groups). This I suggest we can do only if we emphasize to our students that a false theory is oftentimes the foundation of a technology that is more efficient and user-friendly than its more sophisticated, revised versions, and that in certain circumstances, and especially in various practical circumstances, a rough approximation is better and more useful than a refined approximation.

---

[2] Agassi (1987, p. 120) quotes an exceptionally clear example of this kind of conscious blurring of the boundery between collectivism and psychologism in Weber.

Theories that are false because of their simplicity, like the ones discussed above about the physical space of the classroom, can therefore also serve as very useful tools in the hands of journalists, marketers etc.—and can be useful even for the purpose of running a class. Students who apply them report that they acquire a new awareness of the possibility of being themselves controlled or manipulated through the control of their environment, and vice versa: a new awareness of the possibility of minimizing the manners by which they are regularly manipulated by becoming more aware of the ways in which their motives are shaped by control over their environment. I am convinced that this constitutes the most important social value of communication studies, even for those who will not go on to work in the field: communication studies are the best venue for teaching active democratic citizenship in that they train students always to examine critically the possibility of their manipulation as citizens.

So, getting back to the rather minor example that we discussed here at considerable length, one thing I recommend to my own students, whether they are future scholars of communication, future media professionals, or just ordinary citizens, is to choose a different seat in the classroom every time they come to class. As they do so, I suggest that they explore their changing feelings in light of their change of place, as well as any changes their location may trigger in the lecturer and in his behavior toward them, and to note these observations clearly to themselves. Even if the results are not clear-cut and precise, for reasons that we have rehashed, this kind of observation is still highly useful for anyone who wants to enhance awareness of what is taking place around us. This action, especially when constructed as symbolic and applied over and over again in different contexts, and in particular when accompanied by introspection—that is, by a constant attempt to observe what happens to our own psyche and consciousness in light of external changes—gradually transforms more or less naïve students into critical citizens who are aware of the possible influence of context on their own behavior, and of their ability to change contexts actively in light of their own goals and aspirations. This is the whole of the law in a nutshell. Every time we enter a certain space, from our private kitchen to the neighborhood supermarket, we participate in an experiment shaped (not always with sufficient awareness) by the planners of that particular space. The experiment tests the ways in which changes in the environment affect our patterns of behavior. In the case of supermarkets, for example, it has become almost a cliché to point out the size of the cart (which has gradually become grotesque), and the distance within the space of the supermarket between basic items such as bread, dairy products and eggs, as examples of attempts to force us to walk as far as possible among products that we do not need and to feel as if we haven't bought enough. In the case of the supermarket, these features are supported by a frequently evolving technology that monitors how environmental changes incline individuals to act, and how our awareness of these changes alters our behavior patterns. Even the size of the floor tiles can affect how fast we move around the supermarket (more frequent bumping of the wheels of the cart on the tiles causes us to slow down) and thus how much we tend to buy while we are in it, and so on. A sharp communication expert can easily harness even the very superficial generalization we proposed here, along with several similar working assumptions, to an analysis of global communication trends in a way that is illuminating without being dependent on the eternal truth of the theory that underlies his analysis.

Consider the student of advertising, for example. He may benefit from learning, among other things, that the desire to assimilate and the desire for autonomy are the two poles of the famous oral complex according to Freud. Thus we may easily conjecture (assuming, of course, that Freud's theory is correct) that the students who choose to sit at the margins of the classroom will usually tend toward this complex more than their colleagues in the center of the classroom. For the advertiser, this conjecture can be very useful, for instance when he comes to advertise a product that directly caters to oral urges. I don't know how many of my readers still remember cigarette commercials, but they were a distinct expression of the deliberate attempt to hold both ends of their typical group of consumers: the ads expressed a very clear division of the target group into two sub-groups, and encouraged the identification of cigarettes with a satisfaction of both imagined ends of their secret desires—they showed, on the one hand, individuals driven by an extreme and prominent individualistic impulse (with images of solitary, distant journeys on a raft, a horse or motorbike ridden through vast unsettled landscapes, etc.), and on the other hand—social success and manifest assimilation in the social group (for instance, at a cocktail party on a cruise ship, with a group of friends on the beach, etc.).... But let me stop here. As I said, the purpose of this book is not to string together examples that could easily fill the pages of the intro to communication that we are seeking. Rather, its aim is just to clarify the methodological difficulties and principles that ought to guide the authors of that introduction.

Communication studies, as we saw, are concerned with the coordination of orientation challenges and solutions, and its influence upon the conduct of individuals. This kind of study is necessarily multilayered: the coordination (that is to say, communication) between individuals unites them as a meta-system, and the dynamics that drive the new meta-system very soon become a significant part of the environment, the orientation-context, of these individuals, who of course change in light of it. From the communication between individuals, then, emerge new contexts of activity for these individuals, and the new contexts in turn re-shape the individuals in preparation for new problems of orientation.

Bibliography

Agassi, J. (1985). *Technology: Philosophical and social aspects*. Dordrecht: Kluwer.

Agassi, J. (1987). Methodological individualism and institutional individualism. In J. Agassi & I. C. Jarvie (Eds.), *Rationality: The critical view* (pp. 119–150). Dordrecht: Martinus Nijhoff Publishers.

Gellner, E. (1998). *Language and solitude*. Cambridge: Cambridge University Press.

Levine, D. N. (Ed. and Trans.). (1971). Georg Simmel on individuality and social forms. Chicago: The University of Chicago Press.

Chapter 18
Example #2: Stigmergy and Autonomy in the Cyber Age

The example we considered in the previous chapter dealt with the orientation of individuals in their (private) environments and only indirectly with their communication. The students in the classroom influence their environments and are influenced by them on various interrelated levels that are studied by different sciences (despite the fact that they cannot be fully and completely separated). This fact significantly complicates our ability to accurately explain their behavior. And yet, whenever one of these students meets a friend in the cafeteria, when they start talking, when other friends join in the conversation, when they associate and form cultural institutions, when they participate in those institutions and maintain all the many aspects of a communal human life, we have what is of course a far more complex situation. Advanced verbal communication is only possible in the first place because the individuals who partake in it already share a very substantial multi-layered background: biological, cognitive, and cultural. And it is clear that so long as we do not understand how to unite these activities completely by means of reduction, the rigorous refinement and improvement of our theories of verbal communication cannot be completed either. We need to draw the attention of our students, time and time again, to the deceit involved in the idea that there are final rigorous and comprehensive theories of communication. If we get carried away and profess that the theory we proposed (say, universal grammar) somehow captures an essential unit of reality (an innate linguistic organ, for example), if we err in thinking that by virtue of its miraculous ability to capture this essential unit of reality our theory allows us also to give a final and complete explanation of the phenomena we observe (in this case, of the sum of all possible choices available to language speakers in constructing sentences), then we are deceiving ourselves and our students, we become irreversibly embroiled in the blurring of boundaries between theory and research program, between a rigorous formal reduction and the dynamic and elusive reality that we aim to explain. Simply put, *we confuse what science is with what we wish it to be* (as Chomsky and all those who follow him have indeed found themselves doing many times over).

As a second example, which will also serve to sum up the conclusions of this book, I have chosen to present a case that bridges orientation and communication in a helpful way, highlighting the naturalness of the transition between them without diminishing the theoretical difficulties involved in this transition. *I want to*

N. Bar-Am, *In Search of a Simple Introduction to Communication*,
DOI 10.1007/978-3-319-25625-2_18

emphasize the all-important value of reductionism-that-is-aware-of-its-own-limitations, and in particular of the reductionist program when it is contained within a non-reductionist picture of the world. This, after all, was the goal of the present book: not to attack reductionism as a research program, for we have no real substitute for it as such, but to improve it by openly recognizing its limitations. The example that sums up this book allows us also to ascend swiftly from the world of rather simple mechanisms of orientation, the world of termites and ants for example, to the vibrant world of cyberspace and the "smart" metropolis. So that we will now finally touch upon the burning questions of our exciting and intimidating future, and in doing so will also see why the very abstract matters that concerned us thus far are relevant to our understanding of this future.

The example I present here of a reduction of communication and its limitations has very quickly become a classic over the last few decades. It became almost instantly popular thanks to its varied uses as the foundation of a vast number of social networking applications. It goes by many different names in different fields. I will call it "stigmergy", following the famous French biologist Pierre-Paul Grassé, because as far as I know he was the first to provide a distinct description of this mechanism. Machine engineers often know it as "swarm intelligence", while ecologists, evolutionary biologists and immunologists know it also as "collective thinking". And it has many other names in these and other sub-fields. This kind of theoretical fragmentation, which sometimes indicates a substantial waste of intellectual and other resources, stems precisely from the fact that the general principles of communication studies have not yet been articulated in an accessible way, a way that is sufficiently suitable for everyone as a shared foundation for all those various fields. Instead, various researchers have been forced to discover, study and describe anew various aspects of the phenomenon of stigmergy, despite its having already been discovered, studied and described by their colleagues in neighboring fields. Grassé was one of the great zoologists of the twentieth century (and also, incidentally, among his many intriguing credits, a pioneer of neo-Lamarckism, but we won't get into that, although there is a tight and interesting link between his neo-Lamarckism and his views on communication, which I describe below). He sought to solve a fascinating puzzle that has engaged anyone who has ever stood in awe before the dizzying complexity of the coordination displayed by various natural swarms. Termites, for example, are able to build entire networks of dwelling mounds that amount to veritable cities of astonishing complexity, without a master plan or central engineer to guide and direct their action. How does that happen? How does the termite carrying a single grain for building know where to deposit it? Does it understand that it is taking part in the building of an arched ceiling here or an inner hall there, and if not, for clearly it does not, how does it "decide" where to set down the next tiny grain? What guides it, and how? Variants of this problem have for years preoccupied ant researchers as well. It is a well-known fact that given several different and distinct paths to a source of food, for instance, an ant colony—that is, the ants as a group—can "find" the shortest route and stay with it until the food is deconstructed and carried ceremoniously back to their nest. How is this wonder possible? The individual ants, after all, cannot calculate, count steps, or share the results of such calculations with their fellow ants. How, then, can the colony "infer" the shortest route and "convey" this information to its individual members?[1]

By way of solving this puzzle, Grassé described a very interesting form of communication that straddles the thin and bridging line between orientation of the individual within its private environment, and communication between the many within a shared environment or setting. This is also a bridging boundary between mechanical activation (responding to a stimulus) and full-blown communication (the mutual and goal-directed exchange of messages). Grassé gave the particular form of communication he described the name "stigmergy". Unlike the rapid back-and-forth exchange of messages, like the kind that takes place between students in the cafeteria, stigmergy is the delayed and "indirect" activation of a chain (of reactions), a sequential and ordered transmission of orientation challenges. It unfolds as follows: an orienting system, an ant in our case, enters a setting and alters it slightly, in response to some stimulus. The next ant who enters this setting therefore finds a slightly different environment, by virtue of the minute change enacted in the setting by its predecessor, and accordingly, it will respond to its environment slightly differently: it too will cause a minute change of the setting consistent with the same trend. The next ant will also respond accordingly, it too will change the setting slightly and in a way that will affect the environment of the ants that follow, and so forth. Thus, a rolling chain of reactions occurs, a chain of ever-increasing corresponding environmental and behavioral changes: a positive feedback. Finally, the cumulative change reaches a certain threshold (a physical threshold, like the exhaustion of a resource, or a threshold that is also biological, like the neural capacity to carry a certain stimulus, the limit of the processing capabilities of a system etc.). Beyond this threshold, the stimulating action on the agents changes: the positive feedback becomes negative, and the "task" is complete.

Take the intriguing task mentioned above of the ant colony finding the shortest path to a food source. Ants direct each other's action using pheromones: they lay down these action-triggering chemicals along their path, scatter them in the air and even apply them to one another, depending on the context and the task at hand. They even "identify" members of their own colony by means of a pheromone secreted by the queen ant and disseminated with the help of her personal assistants: if we replaced this identifying communal scent in one of the members of the colony, that ant would immediately be identified as an invader and treated as such. And there are also pheromones that signal a warning, those that facilitate an attack, and many more. One well-known pheromone, which we will focus on, can be described

[1]Without getting into too much detail, it is worth noting that treating the individual ant as a distinct biological entity is itself contestable on several counts. With the exception of the queen ant, for instance, ants are all sterile and thus do not pass on their genetic material to offspring. Just like our digestive system, which is so rich in creatures that are not us and without which we could not exist, ants, too, remind us that in nature, the concept of the individual entity is very vague, and that it is only a significant idealization—that is, a conscious limit on our tendency toward reduction—that allows us to render it scientifically useful.

as the "follow me" pheromone, because it directs the ant that detects it on the ground to follow the ants that preceded it, until it finds the next "follow me", which orders it again to proceed (and how): thus, ants create pheromone paths like Hansel and Gretel, or like the trailblazing markers that guide us along hiking paths, and this enables them to march along in their famously long and orderly columns.

Now suppose that three different ants have found three different paths of different lengths to the same source of food, and that they mark their way back to their colony with the "follow me" pheromone (this is a very crude abstraction of what actually takes place, but that's of no consequence to our present interests). What we have here is a classic problem of orientation/communication: How will the ants determine which of the three ants has found the shortest and most efficient path? And how will all the ants be notified which path has been chosen? Grassé's answer to this was both simple and brilliant: at first, an equal number of ants will go down the various paths, marking their path as they go lest it disappear. But soon enough, the shortest path takes on a stronger scent simply because, being shorter, the ants that travel down this path also return faster and walk back again along the path. And because the ants are more powerfully attracted to and activated by the stronger scent, they quickly move to take this path in greater numbers, thereby increasing its scent even more. This positive feedback ensures that within a short span of time the other paths will be largely abandoned. Thus, with no mathematical knowledge and no sophisticated communication, the colony has solved a computational problem.

Now, the first thing to note about this explanation is that it is an exemplary eliminative reduction of what had previously appeared like a mysterious intelligence that emerged out of the blue. (More precisely, it is a quasi-eliminative reduction, as explained in earlier chapters, but we won't dwell on that again.) *Grassé's explanation is therefore an exciting example of successful reductionism*: what was once a scientific enigma concerning a hidden intelligence that emerges from nowhere and appears to require complex communication, is suddenly revealed to be the result of a fairly predictable chemical-biological cybernetic "hardware". And most importantly: *communication, the mutual exchange of messages, has been satisfactorily reduced here to a sum of isolated acts of orientation performed by isolated individuals within their (private) environment.*

The second point worth noting about Grassé's brilliant explanation is the impressive scope of its application to the world of living organisms. Grassé did not solve some marginal problem that had troubled a small group of eccentrics devoted to studying the ways of the termites and the ants; he gave those who came after him a rich explanatory platform that allows scientists to reduce, explain and even control and direct many different cases of otherwise miraculous "coordination" among many different groups of animals: from the incredible acrobatic synchronization of bird flocks in the sky and schools of fish in the sea to the mysterious cases of orchestrated activity we find in plants, fungi, bacteria, and body cells. In each of these cases we are able, to a large extent, to explain and reduce what had previously appeared like inexplicable intelligence or sensitivity, and to do this with fairly simple means. Both the professional and the popular literature of recent years is filled with various accounts of fascinating cases of this sort: *by virtue of stigmergy, we now know, plants appear to "sense", fungi to "remember", the immune system to "learn", and bacterial colonies to "draw conclusions".*

The final point we need to emphasize about Grassé's solution is that *stigmergy is significantly present in the human world as well, indeed it is establishing itself as a particularly prominent mode of communication, one that stands to shape the fate of humanity*—no less. It gained this prominence thanks to the world of the web, as we will soon see, because there, the ability to assemble and mediate the choices performed by very large numbers of individuals and translate them quickly into proposals that affect the behavior of others is unprecedented in its scope and efficiency as *a problem-solving tool.* We ought to focus our attention on the growing uses of this tool in the human world because *it is an exceptionally powerful example of the pair of questions that this book encourages readers to raise anew in every context. The first question is that of the limitations (the advantages and disadvantages) of a strict reduction of our activity (in the case before us, by means of cybernetic-stigmergic models). The second is the question of how best to enhance our awareness to the benefits and risks embodied in every such reduction, in every context, and in every context anew.* In particular, we need to raise this second question over and over again (as it has no single fixed answer) in light of the fact, consoling to some and challenging to others, that unlike ant behavior, human activity is not fully and decisively determined by the stigmergic structures in which we partake, and in light of the further fact, which terrifies many and excites many others, that unlike in the case of ants, human stigmergic structures are nowadays monitored by and accessible to the guiding intervention of individual agents, who are thus able not only to understand them but also ultimately to direct them as they please. By virtue of this fact, these agents become exceedingly powerful, indeed unprecedentedly so. The concentration of this enormous power in the hands of few individuals who have not been chosen for this role is one of the most grave and burning dangers that humanity has ever known. Even if this power rested in the hands of many, or of an elected group, or in the hands of no one at all, it would still pose a potential danger of unprecedented proportion. There is no doubt that we are woefully ill-prepared for this danger, even relative to other dangers that we handle ineptly and irresponsibly, like nuclear proliferation. Thus, the various theoretical possibilities for solving the problem before us all harbor both an unprecedented danger and a great promise for a better future, both personal and collective. Our purpose here is not to determine how we should proceed but just to indicate the vital and urgent importance of a serious discussion of this matter in light of the basic questions we articulated above.

Before turning to consider these questions we should get to know some very simple cases of stigmergy in a human environment that is not at all a cyber environment. Imagine, then, for instance, a person walking for the first time through a vast and dense corn field. He makes his way through the tall stalks in search of his destination. The path he takes allows him and those who follow him not only to advance more easily following the marks that he has left behind, but also, if he returns to the field the following day, to improve upon his first attempt to reach his destination. Suppose he seeks the shortest route to the church that lies at the other end of the field. Over time, it will be easy to observe that a path that does not lead to the church, or one that leads to it with an unnecessary detour, will be covered up again with a thick and tall row of stalks, while the path that was gradually identified as shorter will become easier and more inviting to cross, just as in the case of the ants. The easier the path becomes for walking, the more people will take it, believing of course that it will lead them to a clear destination (the stigmergic model also easily explains how the marking of paths is frequently improved—as a series of miniscule improvements that receive miniscule feedbacks—and not just how a shorter route is chosen among several alternatives). Thus, the quality of the path is an indirect testament to its utility, and the properties of the field are translated here into a platform that people who do not necessarily know each other use, without deliberate intention, to learn about the movement of those who have preceded them and to affect the movement of those who will come after them. They thereby participate, as a group, in a "smart" solution of a problem they all share, without establishing any kind of unifying, mediating institution, that is, without a means of communication that has been deliberately designed to coordinate their mutual expectations.

During my postdoctoral stay in Cambridge, MA, I came across another example that is worth mentioning briefly because it involves not the solution of a simple orientation problem shared by individual agents but the emergence, seemingly ex nihilo, of an impressive artistic expression. I was living at the time near an elementary school whose one side wall had been covered with many thousands of pieces of colorful chewing gum pasted to the wall at more or less equal distances: a spectacularly rich pointillist abstract creation. Someone, probably a bored student on his or her way home, had started pasting their chewed-up gum to the wall every day, and this had soon "infected" many other students with the same action as well as passersby who likely never even met the first student. It was clear that a single individual could not have completed this large-scale project (anyone who attempted this would be sure to get maltitol poisoning), and clear that at least some of the people who had participated in it were strangers to each other (even I contributed one or two colorful units). This was a spontaneous, undirected mass creation.

Today, any child who uses a smartphone or one of the countless "crowd wisdom" websites could give you more up-to-date (but only slightly more complex) examples of this kind of stigmergic intelligence and crowdsourcing. Every navigation application that calculates the best route to your home at rush hour mediates the data from many other drivers as a tool for calculating your most efficient route, and every joke application that finds you new jokes that will likely make you laugh because other users who have laughed at jokes that you have laughed at in the past have also found them funny, is nothing more than a simple imitation of the ants' stigmergic mediation mechanism. The same principle also underlies the software that finds a fashion boutique, restaurant or book to suit your taste based on your own prior evaluations and those of users of similar taste and budget according to data that you and your fellow users have provided in the past. A very similar algorithm matches you up with mates (for romantic purposes or any other purpose, as you specify), whom you will probably like because they have been liked by those whose documented choices thus far have resembled yours. And so forth.

It is not surprising, then, that stigmergy has been enjoying unprecedented popularity in the last two decades: it has become the ultimate path of web entrepreneurs to meteoric wealth, with minimal investment and at a rate that has no parallel in human history. The common idea of all these projects is incredibly simple: offer the crowds a mediating platform for communication, one that will allow them—casually and as the by-product of some other action—to document their choices, while as a group they solve a problem that concerns each one of them and which no single individual among them stands a chance of solving on their own (within a reasonable timeframe); quantify the "crowd wisdom" accumulated through the stigmergic mechanism, and solve the problem for them. Highly promising models of this sort are recently being proposed as tools for the more efficient management of big cities, for example: for improving transportation, sanitation, water and electricity consumption, and even improving estimates about the spread of epidemics. Using these models, the number of buses allotted every line, for instance, is determined not just on the basis of vague conjectures regarding traffic patterns in light of past statistics but rather in real time by crowdsourcing the GPS information from phone users at the various bus stops and on the buses, information about actual traffic congestions on every given road at every given moment (the latter also handled through the stigmergic control of traffic lights etc.,). These methods both improve service for the customers and reduce costs for providers, a rare combination in many other contexts.

The fast, big profits that these methods guarantee often result from a seemingly indirect and innocuous aspect of their use: while stigmergic application softwares collect the data necessary for their operation, the designers of these applications also accumulate statistical information of incredible scope and accuracy about our habits of behavior, the habits of response, choice, and movement of each and every one of us, information about the ways in which our choices change in light of various changes in the contexts we are in, and more. Already today this statistical data can be crosschecked against other similar databases to create a model of unprecedented scope and accuracy of some of our behaviors as a swarm. And also, if the need arises, of the behavior of each and every one of us as individuals. In many contexts (and their numbers are growing at a dizzying speed), this information becomes a fairly reliable tool for simulating our behavior patterns, our personality, and in some cases—and this is of course the most frightening danger and fascinating promise of all—it is already used to direct our behavior and to fine-tune our orientation patterns both as a swarm and as individuals. After all, a very large part of our actions in the world is predictable, and this fact is unsurprising: it represents a crucial form of conservation of biological resources, without which life cannot exist. Even what we call, rather vaguely, "a person's character" is, as we know, largely a hardening of his or her many patterns of response in similar contexts. Thus, just as we tend to choose more or less the same seat in the classroom every time, so we tend to have rather fixed patterns of behavior when we are anxious or sexually aroused, when we play, when we are in love, when we tackle problems (of a similar nature), and of course also (and in particular) in our web behavior. Now, clearly, when the accumulated information about these patterns encompasses our shopping habits, it becomes highly valuable for those who have the narrow marketing interest. But when it also encompasses, and even if only for marketing purposes, our hobbies and areas of specialty, our expected location one minute from now and our probable location tomorrow, as well as our intimate preferences and details about our health condition, including our heart rate and blood pressure in the presence of various stimuli, when it encompasses so many aspects of our personality, which is rather fixed, it seems there is no parallel in all of human history for information that is at once so dangerous and so promising. This information is constantly being collected and constantly becoming more specific and more accurate thanks to our growing use of stigmergic algorithms, and all this at what is considered a completely negligible cost in the realms of the big data-collection and crowdsourcing centers, which offer us most of their services free of charge, of course. And indeed, as many before me have noted, nothing sums up this alarming and exciting situation better than the famous quip that if you're not paying for the product, you are the product. (This is not true of algo-trading which deserves a separate chapter and will therefore not be discussed here, although of course it is an important recent case of stigmergy in the service of the few).

This frighteningly vast scope of information mining by means of stigmergic algorithms is also at the heart of the deeply confused debate about the possibility of "artificial intelligence" that exceeds the intelligence of the humans who feed it. Since all these mediation mechanisms (stigmergic and other) simulate "sensitivity" and "wisdom" and of course also an "intimate acquaintance" with our preferences, it is very easy to confuse this with actual sensitivity, wisdom and intimate acquaintance. It is perhaps worth stressing that many of our preferences, which are now exposed through the mechanisms that monitor our habits, were fairly hidden until recently, even from our own selves, and this is a big reason for the great and justified excitement here. Our preferences were until recently not only private but also unknown: most of them were unknown even to us, because never before had we come face to face with such a clear, detailed, unbiased and unsparing reflection of the patterns of our own behavior. After all, not all of us are eager to learn just how predictable we are, not all of us are eager to admit even to ourselves everything that stirs us, everything that frightens us, etc. So that this development is very exciting. Yet these mediation mechanisms, stigmergic and other, are of course not in the least bit intelligent (just as the ant colony does not really have a will or a brain, to say nothing of self-awareness). They are just gradually improving reflections of *our* behavior as these are returned to us by constantly improving technological means. Time and again, lay people are surprised by the miraculous "wisdom" of the search engine, for example: it appears to be able to "understand" our intentions and desires and to "find" just what we are looking for. But the wisdom here does not belong to the algorithm of the search engine, which is of course not even literate, just like the wisdom of finding the shortest route to the church at the far end of the field does not belong to the field or to the path that points the way toward it. The "wisdom of the web" lies always and before all else with the users themselves, as individuals, and in the brilliant simplicity of the mode of communication that mediates for them an unbiased mirror of their collective activity: the secret lies always in communication. We tend to form habits, we respond in similar ways to similar environments, and often we are willing to conduct ourselves within environments that allow for a substantial narrowing-down and rigidification of our response patterns. Under these conditions, mediating our choices and quantifying them (stigmergically or otherwise) is not an overly complicated task and it requires no miraculous understanding, to say nothing of wisdom or consciousness.

Consider again the miraculous "wisdom" of the search engine. It depends above all on the computer's most conspicuous and least interesting advantage over the human brain: its great computing capability, which is here translated, *inter alia*, into an unprecedented ability to map the sites on the web according to the presence of keywords, their combinations with other words, etc. All this, of course, without actually understanding the first thing about the content and meaning of what users say (as we saw in the chapter on Shannon). A supercomputer can scan a dazzling number of web pages in a fraction of a second and search for combinations of various "clue" words that provide (for humans) a highly probable indication of the content of the site. This data can then be unified, enhanced and improved by means of various stigmergic mechanisms that increase their accuracy: data about the prominence of websites according to the number of inbound links (links from other websites to those sites), about the level of traffic to and from the sites and the number of individual entrances to the sites, the satisfaction of end-users who used specific keywords that lead to them, the presence of other keywords typically searched by users who have searched similar keywords, and so on. And of course, each one of these methods and many more will continue to improve at a dizzying speed. (By the way, like every positive feedback, this data functions a bit like a self-fulfilling prophecy: for example, we rarely look beyond the first ten results that the search engine offers us, and thus we further reinforce that search engine's suggestions for the next user; similarly, by preferring a particular search engine we actually improve it and help it outdo its competitors.) And in the background of all this there is of course also data about various searches by people who are like us (in relevant ways) that have ended with a purchase. Thus, just like the chosen path announces itself to the ants through its more powerful scent, the search engine, and this is no secret, aims to offer users constantly improving ways of finding what they are looking for and at the same time to provide advertisers with an increasingly accurate segmentation of the group of users at which they should target their product. This segmentation has many advantages, not all of them malicious or even negative (Vaidhyanathan 2011, pp. 28–30) : a sufficiently accurate segmentation (and especially one that incorporates data about our location, hobbies and interests, birthdays, etc.) undoubtedly helps small businesses compete successfully with big ones in the public auctioning of "banners" that a search engine like Google conducts with every single search, an auction whose immediate results appear alongside and above the regular search results as "sponsored links". If the segmentation applied to the user is accurate enough, even owners of small business with modest resources will know exactly when it is worth their while to spend more money than their richer competitors in order to upstage them at a critical moment (of course, they do not need to "know" anything in real time: the stigmergic mechanism "knows" this for them and is supposed to purchase the desired ad accordingly). We all benefit from this, then—even if the owners of the search engine benefit a lot more than everyone else—because regardless of whether or not a commercial transaction takes place, the model of our behavior patterns recorded by the search engine constantly improves (this is the real-life embodiment of Warren Weaver's dream).

Let us focus now on several of the most prominent dangers embodied in this process. Some of them are related to the fact that the vast amount of personal information collected about us, and through us, does not really belong to us: all too often we are not exposed to it and cannot use it at will. This is one of the greatest failures of our current social organization, and it testifies above all to the pathetic way in which the legal, legislative and governing systems lag behind our technological and commercial development. And of course, many other dangers follow from this state of affairs: our personal information is now only a short step away from the manipulative management of those who run the stigmergic crowdsourcing systems. No "big brother" or "responsible adult" oversees the ant colony, but here a person or group of people whom we have not chosen and about whom we know next to nothing are in charge. We can only hope that they are legally adult and emotionally mature. Those who preside over the stigmergic system can, for instance, have access to our medical records, delete access to competing businesses, change our driving routs and alter our personal information without our knowledge, and this is perhaps the least of their abilities to harm us. A frightening example of such manipulative possibilities was provided recently by Facebook, which markets itself to the public as a free social network and to advertisers as a mechanism of unprecedented efficiency for profiling, monitoring and segmenting customers. Facebook representatives, we all read in our daily newspapers, boasted that the company offered different sample groups of users a virtual "I Voted" button with which they could declare their participation in the US presidential elections, and argued that the size and location of the button ultimately had significant impact on the voting rates of users (it seems almost needless to say that, if all this is true, these users participated without their knowledge or consent in an experiment in manipulation that is at the very least intrusive and possibly also illegal). Supposing for the moment that the experiment was properly designed and legally carried out, and that its conclusions are reliable, this means that in the case of a close race, and given the existence of prior knowledge about the users' political leanings (a very simple matter for the enormous data-collection mechanism that is Facebook), Zuckerberg and his friends can tilt voter turnout among a certain relevant segment and thus decide the elections. Just like that. And what is perhaps even more frightening (and exciting), as well as indisputable, is that this constitutes just one of many examples of the ways in which we are used as research animals—without our knowledge, to say nothing of our consent. Our sexual orientation is correlated with data about our dietary habits by websites that presume to provide us with dating services; our tendency to have intimate relations and even to cheat on our partners is predicted by algorithms that track changes in our vocabulary, etc. And all in the name of improving service. There is no limit to the interesting segmentations we could choose to pursue here. The problem is that the experiment is not public and the subjects are not participating of their own accord. We gained free membership to a social network, and have thereby unknowingly become lab rats in the most impressive and comprehensive psychological laboratory in human history. And the knowledge we provide, the real gold of this world, lies in the hands of those who gave us the new mode of our communication, with apparent utopian generosity.

It is worth emphasizing that a more effective and more promising tool for self-development than an accurate and unbiased reflection of our behavioral patterns is hard to imagine. And the greater the share of our activity that takes place online and is recorded, the sharper the mirror becomes and the more profound become the ways in which it can potentially help us. Suppose, for instance, that tomorrow we all receive, free of charge, wrist watches that can measure our blood pressure and heart rate (today, mostly due to the shortsightedness of marketers and the blindness of consumers, these devices are sold as expensive accessories for flashy gadget buffs, but I have no doubt that before too long they will be given out for free). Whoever is able to correlate the information that accumulates in this mobile monitoring device, this polygraph, with the contents we consume can provide us with many valuable conjectures about our psychic structure, obviously, about our behavior, our health for instance, and about ways to improve it. They can offer us conjectures about what causes us to be stressed and how better to relax, what we are unconsciously attracted to and what secretly repulses us, can tell us when we are tending toward depression before we even sense this as well as how to respond, how to improve our eating habits and when to pay them particular attention, when to rest and when to work out, and on and on. Obviously, this is a powerful tool for personal development, an unprecedented expansion and improvement of current biofeedback methods (whose guiding spirit is of course Norbert Wiener, the same Wiener whose writings are still not taught in the philosophy departments I attended). This information is already used by data-mining companies to form a comprehensive model of our habits and of the ability to reshape them, of the ability to influence our inner states and our behavior. It is a model that in many cases stands to save lives and to improve their quality, of course, and in other cases may be used to destroy them or even to cause us to destroy them. Another current initiative proposes to unite our genetic information and much of our epigenetic information and to join them to our medical history as part of a stigmergic mediation algorithm that will offer us the probabilities of our tendency to develop certain diseases in light of certain environments, and even to recommend how and where we ought to live so as to avoid them. This is not science fiction: it is already happening today on a small scale, and is already (with some justification) no more than a luxury toy for the rich. The great biologist Richard Lewontin drew this out well when he pointed out that if increasing the lifespan and improving the quality of life of the general population were foremost on the minds of those who launch these initiatives, these goals could have been achieved through trivial and much less costly means, like improving air and water quality (Lewontin 1993). Who, then, ought to have access to this detailed information? And how should we supervise those who have it? Can we possibility allow such powerful information to be held by a common money-monger or even a good wishing entrepreneur? Should insurance companies, for instance, have access to it? Should it be held by our doctors? Our governments? By us as individuals? Is it even possible to prevent the leaking of this sort of information and its use by those who have interests that are foreign to our own? These are all highly urgent questions (in a sense, they are already outdated questions, since a vast amount of that information has already been amassed and analyzed). *It seems clear that as a society we are as yet unprepared for the incredible range of influences of our new mode of communication, and herein lies the problem.*

I have made a point, above, of using only well-known examples and ones that have already featured in the popular and quasi-scientific media. Our interest here is not to expose new facts and certainly not to decide the important debate about our desired political and economic future and about the proper supervision of our new modes of communication in the utterly new circumstances that we face. That discussion deserves much more than a separate book, let alone a final illustrative appendix whose goal is entirely different. My aim in this chapter is just to point out that the proper focus of the vital public debate proposed here has already been placed at the center of this book: we need to carry out a meticulous survey of the achievements and the limitations of reducing this new-old mode of communication, and in particular to examine the reliability of such a reduction as a tool for describing what is actually taking place. In other words, we need to appreciate the significant difference between an ant colony on the one hand, and a mass of people stigmergically mediating information in an attempt to improve their problem-solving abilities on the other hand, and to try and re-organize our society in light of our conclusions. In particular, we need to be prepared to change our institutions and our behavior in light of what we learn about the limitations of stigmergy as a theory for managing society and the individual. For example, *we ought to posit, with great caution, a question that to the best of my understanding no supercomputer comes close to answering in a way that would be counted even as a reasonable imitation of intelligence, namely: What characterizes all the problems whose solution will not be found in stigmergic models? And what characterizes those contexts in which stigmergy is manifestly preferable?*

A complete and rigorous answer to this question probably does not exist. This, in any case, is a conclusion that follows from the basic claim that this book has advanced: the new tools we develop to solve the old problems change us in ways that are sometimes unpredictable, and with us evolve also the old problems and contexts. But *the most prominent advantage of stigmergy as a tool for managing and advising both the individual and society seems to be grounded in contexts in which we are willing to accept a certain uniformity and a restriction of our creativity, that is, in rather rigid and limited choice environments.* So long as the problem we face arises within environments of this kind, environments in which our likeness to others is desirable, stigmergy stands to provide us with an excellent solution. In other, more subtle choice environments, the great power of stigmergy and especially its total blindness to the changing context is a double-edged sword. For, the stigmergic mechanism is undoubtedly not just faster and more efficient than all the greatest human minds put together, it is also "dumber" than the dumbest end-users who feed its superhuman "wisdom". An ant cannot willfully resist the stigmergic dynamic in which it participates, and this is why Grassé's proposed reduction is near complete. In the human case, as we have seen, this is not so: having cracked the secret of the ants, for example, human beings can now easily lead them to the horizon or round and round in circles till they drop, and in just the same way, a person can try to turn his friends into a community of obedient ants. But humans may also choose to refuse the guidance of the system, be it a social context or a search engine, for no special reason—other than perhaps the sheer joy of refusing. They may, for example, divert the stigmergic mechanism's direction to the point of absurdity, as various self-described freedom fighters occasionally do. (They can cause a search engine, for example, to suggest pictures of a hamster as the results of a search for the "God particle", or to encourage marketers in different ways to sell ice to Eskimos). In principle, these scenarios are no different from the ability of travelers in the corn field to choose to clear and conquer paths that lead nowhere, thereby misleading those who follow them. And of course, we may in some cases bring about a trend that is preferable to the one that had emerged from the system. *The value of this new and original solution will rise the more the uniformity produced and facilitated by the stigmergic system grows. Thus, the incentive for an original, non-stigmergic solution increases the more the system's efficiency grows, and this new solution will of course also reshape the system's direction.* Only the context-sensitive human being will be able to evaluate such original solutions.

The advantage of stigmergy as a tool for solving problems, then, stands out especially in structured choice environments, with their pre-given and clearly distinguished alternatives. It diminishes the more the problems to which we apply it require creativity, sensitivity and originality. The programmers of the stigmergic algorithms are well aware of this, of course, which is why one of their main goals is to rigidify and simplify our choice environments as part of the operating systems and the interfaces they offer us, while they themselves retain the status of those who see without being seen and critically evaluate without being criticized. It is important to emphasize this point in summing up because we really ought to note the fact that in their possibly innocent attempts to improve the efficiency of their stigmergic systems, many of our "big brothers" have become the enemies of human progress. For their purpose as a business is to rigidify our behavior and confine it as far as possible to a fully definable frame of action based upon pre-given and fixed categories. But if McLuhan was right, and on this count I have no doubt that he was, if the modes of our communication reshape our personalities and our patterns of behavior, then our growing use of such interfaces risks making us simpler and more predictable, less original and less creative. *In their effort to make the models that predict our behavior ever more accurate, the aim of these "big brothers" is congruent with the effort to "dumb down" end-users, or make them shallower.* The paradigms here are the famous "thumbs-up" and "thumbs down" buttons, which divide the entire universe with one dim thrust of a sword into votes of likes and dislikes free of all nuance or explanations. But the roots of the problem, as many have noted, go all the way back to the pre-fixed menus of the operating systems of the old Apple computers (a system known today as Microsoft's "Windows"), and maybe even, as Jaron Lanier recently argued, back to the rigid data structure of the "files" system that supports and enables these operating systems. I am convinced, for instance, that as a society, in preferring these rigid operating systems over open-source systems we are encouraging the dumbing down of our society. Aristotle's essentialist vision is becoming true before our eyes, but not because it succeeds in capturing the true nature of reality, which is always illusive; rather, because we have succeeded in creating for ourselves a rigid artificial environment made of more or less uniform Lego blocks, and this reality in turn is shaping us as less creative and much more predictable creatures.

Communication, say the essentialists, is the mutual exchange of messages. Let us grant them this contention. Still, the message we receive from our interlocutor is a *message*, and not a purely mechanical activation, because as an attempt to activate us mechanically, its force has been blunted by another mechanism, a supervising mechanism, which is context-sensitive and not always predictable. *The message is a message because unlike in the case of the mechanical activation, we can resist it, and because this resistance is context-sensitive, and sometimes (no matter how rarely), intelligent and creative.* Since contexts and the movement between them cannot be fully and rigorously quantified, there is and will be no rigorous and complete theory of communication, at least so long as the basic rules of the extensional game remain unchanged. Yet because the constant improvement of monitoring technologies is accelerating at such dizzying speed in recent years, it seems to many people that the rules of the game have indeed changed. It is therefore important to emphasize that the rules of the game have not changed, and that to the best of our understanding, they will only change for the worse if we accept the gradual rigidification of our choice frames. This is, if you will, the greatest danger that we face today. And our task as communication students is to ward it off.

Bibliography

Grassé, P.-P. (1982–6). Termitologia. 3 vols. Masson, Paris

Lanier, J. (2006-05-30). Digital maoism: The hazards of the new online collectivism. *Edge*

Lewontin, R. C. (1993). *Biology as ideology: The doctrine of DNA*. New York: Harper Perennial.

Sassower, R., & Bar-Am, N. (2014). Systems heuristics and digital culture. In D.P. Arnold & R. King (Eds.), *Traditions of systems theory: Major figures and developments* (pp. 277–292). London: Routledge.

Vaidhyanathan, S. (2011). *The googlization of everything*. Berkeley, LA: University of California Press.

Name Index

N. Bar-Am, *In Search of a Simple Introduction to Communication*,
DOI 10.1007/978-3-319-25625-2

Subject Index

N. Bar-Am, *In Search of a Simple Introduction to Communication*,
DOI 10.1007/978-3-319-25625-2

Printed in the United States
By Bookmasters